NISTIR 7383

Selected Procedures for Volumetric Calibrations (2012 Ed)

Harris, Georgia L.
Weights and Measures Division

March 2012

U.S. Department of Commerce
John E. Bryson, Secretary

National Institute of Standards and Technology
Patrick D. Gallagher, Under Secretary of Commerce for Standards and Technology and Director

Foreword

This NIST IR of Selected Publications has been updated from the 2006 version and includes Good Laboratory Practices, Good Measurement Practices, and Standard Operating Procedures for volumetric calibrations.

Many of these procedures are updates to procedures that were originally published in NBS Handbook 145, Handbook for the Quality Assurance of Metrological Measurements, in 1986, by Henry V. Oppermann and John K. Taylor. The 2006 updates incorporated many of the requirements noted for procedures in ISO Guide 25, ANSI/NCSL Z 540-1-1994, and ISO/IEC 17025 laboratory [quality] management systems. The 2012 updates include updated terminology consistent with the International Vocabulary of Metrology (VIM), primarily associated with metrological traceability. The 2006 changes incorporated 1) uncertainty analyses that comply with current international methods in the Guide to the Expression of Uncertainty in Measurements (GUM) and 2) measurement assurance techniques using check standards. No substantive changes were made to core measurement processes or equations in the 2012 edition.

Updates to this edition include the latest CIPM accepted equations for the calculation of water density in GLP 10. SOP 17 and 20 have been modified so that they apply to control charts for any check standard and range charts and standard deviation charts for any process (versus only those for volumetric calibrations as previously written.)

The following Practices and Procedures that were included in the 2006 edition have been *removed* in this publication:

Standard Operating Procedures for:
- Gravimetric Volume Calibrations Using a Single-Pan Balance (13) and Using Equal Arm Balances (15). Contents specific to glassware calibrations have been included in SOP 14. Few laboratories maintain mechanical balances or equal arm balances for gravimetric calibrations.
- Small Volume Prover, Water Draw (26) and Small Volume Prover, Gravimetric Calibration (SVP). These two procedures are undergoing major revisions to consider additional technologies and will be updated at a later pending review by designated working groups.

Special thanks go to the following individuals for the critical editorial reviews performed for the 2006 edition:
- Kelley Larson, AZ Department of Weights and Measures (retired as of 2012)
- Dan Newcombe, ME Department of Agriculture Metrology Laboratory
- William Erickson, MI Department of Agriculture, E.C. Heffron Metrology Laboratory (retired as of 2012)
- Craig VanBuren, MI Department of Agriculture, E.C. Heffron Metrology Laboratory
- Van Hyder, NC Department of Agriculture & Consumer Services, Standards Laboratory

- L.F. Eason, NC Department of Agriculture & Consumer Services, Standards Laboratory (retired as of 2012)
- Val Miller, NIST, Office of Weights and Measures
- Carol Hockert, NIST, Office of Weights and Measures
- John Wright, NIST, Process Measurements Division
- Dan Wright, WA State Department of Agriculture (WSDA), Metrology Laboratory

Note regarding SI units: This publication conforms to the concept of primary use of the International System of Units (SI) recommended in the Omnibus Trade and Competitiveness Act of 1988 by citing SI units before customary units where both units appear together and placing separate sections containing requirements for SI units before corresponding sections containing requirements for customary units. In some cases, however, laboratory standards and/or trade practice are currently restricted to the use of customary units; therefore, some procedures in this publication will specify only customary units. Non-SI units are predominately in common use in State legal metrology laboratories, and/or the petroleum industry for many volumetric measurements, therefore non-SI units have been used to reflect the practical needs of the laboratories performing these measurements as appropriate. SI units are used where practical, and where use or potential conversion errors, will not likely impact the quality of laboratory measurement results.

Endorsement Disclaimer
Certain commercial equipment, instruments, or materials are identified in this publication in order to specify the experimental procedure adequately. Such identification is not intended to imply recommendation or endorsement by the National Institute of Standards and Technology, nor is it intended to imply that the materials or equipment identified are necessarily the best available for the purpose.

Table of Contents

This page is blank.

Good Laboratory Practices

GLP 10

Good Laboratory Practice
for the
Purity of Water

Water is used in two ways in the metrology laboratory. It may be used as a cleaning fluid or it may be used as a standard of volume for purposes of calibration. In each case, it must be clean; in the latter case it must be pure, as well.

Cleanliness of water may be achieved by removal of physical contaminating substances, especially by filtration. City water is ordinarily clean but may become dirty from the distribution system and especially from prolonged standing in some kinds of pipes and tubing. Hoses used to transfer water from and into large vessels and tanks may need internal cleaning, as well. Flushing to remove visible contamination is all that is usually required.

Clean water is all that is necessary when making measurements by volumetric transfer since only volumetric comparisons are involved. When a water density value is used in an equation for volume transfer, the density values for the standard and unknown are proportional to each other; therefore, deviations in the density values from pure water are insignificant. (This is *not* true for gravimetric calibrations.) The coefficient of expansion for distilled water and pure water are also essentially the same. It is obvious that dirty water could cause a number of problems, including depositing residues in a volumetric vessel that could cause volumetric errors or could soil its interior, as a minimum. When in doubt of the cleanliness of the water supply, filters should be attached to the supply lines used.

Pure water is needed for gravimetric volume calibrations. For gravimetric calibrations, filtration systems alone are *not* adequate. Pure water requires the removal of chemical contaminants and this may be achieved by distillation, reverse osmosis, ion exchange systems, or combinations of these systems. The purity of the water from any given system and maintenance requirements is often dependent on the quality of the source water. Source water should be tested to determine the best type of system to meet laboratory needs. Density calculations may be accurate even when appreciable levels of dissolved salts remain in the water. However, since it is very difficult to know what type and quantity of salts are tolerable before density is affected, it is prudent to use the best system practical within budgetary constraints.

ASTM Type III or IV Reagent Water[1]* is recommended as adequate for gravimetric calibration purposes. Such water may be produced by distillation or by ion exchange with relatively inexpensive equipment. In fact, most systems designed to meet these specifications actually provide quality better than required. Type III grade of reagent water may be prepared by distillation, ion exchange, continuous electrodeionization reverse osmosis, or a combination thereof, followed by polishing with a 0.45 μm membrane filter. Type IV grade of reagent water may be prepared by distillation, ion exchange, continuous electrodeionization reverse osmosis,

---

[1] ASTM D 1193 - (06)2011, Standard Specification for REAGENT WATER, ASTM, http://www.astm.org, 1916 Race St., Philadelphia, PA 19103.

electrodialysis, or a combination thereof. Sales literature will usually specify whether the equipment will provide water of the above quality. Additionally, many manufacturers will provide a source water quality test and recommend a system to meet the purity and volume requirements of the laboratory. There are a number of commercial sources for such equipment.

A cartridge-type ion-exchange system is recommended for its simplicity and ease of operation. It can operate intermittently (on demand) and requires little or no maintenance except for change of cartridges, the need for which will be indicated. A relatively small system (2 L/hr to 30 L/hr) is adequate for laboratories calibrating glassware and test measures up to and including 20 L (5 gal) standards. It may be used on demand or to fill a small (20 L to 40 L) storage bottle to assure a continuous supply of calibration water.

There are two broad types of ion-exchange systems. Pressure cartridge systems (PCS) operate directly from line pressure (up to 700 kPa) and need no special operation precautions. The less expensive type operates from the water line through a needle valve to produce a specified flow rate through the cartridge. In this system, the outlet must not become blocked or turned off to prevent the water pressure from building up and bursting the cartridge. It is common practice to plumb directly from the output of this cartridge to a storage tank without using a valve in between. The unit is operated by simply turning the shut-off valve located at the water supply.

Water density may be measured with a five- or six-place oscillation-type density meter calibrated using suitable standard reference materials that are representative of the range of use. Less accurate density meters are not suitable for evaluating the quality of water needed for gravimetric calibrations. These systems typically measure the density at a specific reference temperature (generally 20 °C). Density meters are generally not needed if an appropriate water purification system is used that includes a way to measure conductivity or resistivity.

Conductivity and resistivity measurements do not have a direct correlation to water density, which is the critical attribute of concern for gravimetric volume calibrations. However, conductivity or resistivity measurements are a good indicator of water quality and whether the system is in good operating condition or needs service. Conductivity is simply the reciprocal of resistivity. For water quality specifications and assessment purposes, conductivity is usually measured in microSiemens per centimeter (μS/cm) and resistivity is usually measured in megaohms-centimeter (MΩ·cm), both usually at a reference temperature of 25 °C. Conductivity is greatly influenced by temperature and is not linear. However, this is not a major concern for typical laboratory applications requiring pure water. Either a meter or indicator light should be included in laboratory systems to monitor water quality output. Standard reference materials are available to test conductivity and resistivity units. But, because the measurement values are not used to perform corrections to volume calibrations, traceable calibrations of the units are not essential.

Exposure of pure water in storage to air will likely cause degradation in the conductivity and/or resistivity measurements. However, pure water has been stored for over a year with little degradation in the density quality (provided that storage containers and lines are clean and that there is no bacterial growth, algae, or other contamination).

The ASTM D-1193 specifications for conductivity and resistivity are noted in Table 1.

Table 1. Conductivity and resistivity specifications for water.

	Type I	Type II	Type III	Type IV
Electrical conductivity, max, $\mu S/cm$ at 298 K (25 °C)	0.056	1.0	0.25	5.0
Electrical resistivity, min, $M\Omega \cdot cm$ at 298 K (25 °C)	18	1.0	4.0	0.20

Conductivity and resistivity (along with other water quality measurements) are often used to assess the water quality used in cooling towers, boilers, relative humidity systems, micro and nanoelectronic systems and in pharmaceuticals, to ensure water of sufficient purity and to minimize corrosion or build-up within such systems.

Temperature equilibrium is another important factor in density stability of water. This is especially critical for large volumes. Water temperatures must be stable. Temperature accuracy is as important as purity for a correct density determination. If water is coming straight from the tap through the purification system into the prover, the temperature may fluctuate appreciably. Therefore, it is important to store an adequate volume of water to complete a calibration either already purified or ready to go through the system.

Water density tables (see Table 9.8 in NISTIR 6969 or Handbook 145) or calculations are used in most gravimetric calculations. For use in computer programs (most often spreadsheets), the use of a calculation is often preferred to look-up tables. The following equation is recommended for use.[2]

$$\rho(t_w) = a_5 \left[1 - \frac{\left((t + a_1)^2 (t + a_2)\right)}{a_3 (t + a_4)} \right] \qquad \text{Eqn. 1}$$

where:
$a_1 = -3.983\,035$ °C
$a_2 = 301.797$ °C
$a_3 = 522528.9$ °C^2
$a_4 = 69.348\,81$ °C^1
$a_5 = 999.974\,950$ kg m^{-3}
t_w is the temperature of the water in °C.

In Excel Format:
=999.97495*(1-(((G8-3.983035)^2*(G8+301.797))/(522528.9*(G8+69.34881))))

where, G8 is the cell with the Celsius temperature and units are given in kg/m^3.

---

[2] M. Tanaka, G. Girard, R. Davis, A. Peuto, and N. Bignell, Recommended table for the density of water between 0 °C and 40 °C based on recent experimental reports, Metrologia, 38, 301-309 (2001).

To adjust the air-free water density in Equation 1 beween 0 °C and 25 °C to air-saturated water (the standard laboratory condition), use the following equation,

$$\Delta\rho\,/\,kg\ m^{-3} = s_0 + s_1 t \qquad\qquad \text{Eqn. 2}$$

where,

$S_0/(10^{-3}\ kg\ m^{-3})= {}^-4.612$
and
$S_1/(10^{-3}\ kg\ m^{-3}\ °C^{-1})= 0.106$.

In Excel Format:

=(-4.612+0.106*G8)

where, G8 is the cell with the Celsius temperature. (Add this value to the density. Air saturated water will be less dense than air free water.)

Note: Equation 1 provides water density in kg/m^3. To convert the result to units of g/cm^3 or g/mL, divide the result by 1000. When adjusting the value for air-saturated conditions, use Equation 2 before converting to other units. The result in Equation 2 will be in "parts per million," so divide by 1 000 000 to find the change to the water density in g/cm^3.

Additional References:

ASTM D 1125 – (95)2009, Standard Test Method for Electrical Conductivity and Resistivity of Water, ASTM, http://www.astm.org, 1916 Race St., Philadelphia, PA 19103.

ISO 15212-1:2002, Oscillation-type density meters – Part 1: Laboratory instruments, ISO, 2002.

GLP 13

Good Laboratory Practice
for
Drying "To Contain" Volume Standards

Vessels calibrated "to contain" must be dried of all measurable water in order to obtain an "empty" weight. The drying process should not contaminate the container; otherwise, it will need to be re-cleaned before further calibration or volumetric use. The following is presented as guidance when drying is required.

Drain as much of the residual water as practical before starting any drying process. If time is not a consideration, a glass tube may be inserted into the container to pass clean dry air (or nitrogen) through it to evaporate the residual water film. A filter or dust trap may be necessary to pre-clean the air used. Air lines must be selected to ensure that aging and wear do not introduce contaminants. Air may be sucked through a tube connected to a vacuum pump with some danger of drawing in dirty air from the surroundings. Clean absorbent cotton placed at the neck opening can minimize the entrance of foreign matter. Compressed air systems may introduce finely atomized oils or moisture into the air which may not be visible. If laboratory quality air is not available, nitrogen may be used.

Solvent cleaning may be used with alcohol as the preferred medium. Preliminary rinsing with acetone will remove large amounts of water, with which it is infinitely miscible, but this solvent often contains impurities such as traces of oils that could deposit on the container walls. Thus, a final alcohol (preferably ethanol or methanol) rinse is recommended, even if acetone is used to remove most of the water. Care must be taken to ensure that the alcohol is not denatured with oils to an extent that will leave residue on the standard. Care must also be taken to avoid mixing chemicals such as acetone and alcohols. The alcohol is allowed to drain as much as possible, followed by air drying as described above.

Some metal containers have been internally coated to minimize corrosion. When present, it should be ascertained that such coatings are not affected by alcohol or acetone; otherwise, solvent treatments should not be used.

The external surfaces of all containers should be clean when gravimetric calibration is used. Otherwise, any removal of external dust or dirt during the measurement process could cause errors of unknown magnitude.

Analytical glassware should not be dried by heating in an oven as the glass may suffer non-elastic expansion and put the glassware out of calibration.

Calibration Tip: After the vessel is clean and dry, cover to minimize contaminants from collecting inside the vessel. Clean and dry the flask one to two days prior to the calibration and allow the flask to come to equilibrium with the environment. Obtain an initial baseline "dry weight" of the clean container for use as a dry reference weight on subsequent weighing and drying cycles.

Safety Note: Material Safety Data Sheets (MSDSs) must be available in the laboratory and should be reviewed to ensure safe handling of all chemicals noted in this procedure.

Good Measurement Practices

GMP 3

Good Measurement Practice
for
Method of Reading A Meniscus Using Water or Other Wetting Liquid

Two common methods are used for setting a meniscus. The method used in calibration should be consistent with the intended use of the volumetric standard. For interlaboratory comparisons, the method to be used should be defined during the planning stages of the comparison.

In all apparatus where the volume is defined by a concave meniscus, the reading or setting is made on the lowest point of the meniscus. In order that the lowest point may be observed, it is necessary to place a shade of some dark material immediately below the meniscus, which renders the profile of the meniscus dark and clearly visible against a light background. A convenient device for this purpose is a collar-shaped section of thick black rubber tubing, cut open at one side and of such size as to clasp the tube firmly.

Two common types of meniscus readers are available. These include black/white meniscus card readers or magnifying glasses with cross-hairs. The width of the graduation will affect the readability of the meniscus and should be estimated to the nearest 1/10 of a division. Black/ white meniscus card readers are simple to create as shown in the figure below. More elaborate card readers may be purchased that allow placement around the neck of a flask.

This type of meniscus reader is generally preferred for Option A method of reading the meniscus. Another type of meniscus reader consists of a clear lens (plastic or glass) with etched lines on the front and back that are aligned to prevent parallax errors in reading. This type of reader is preferred for Option B.

The meniscus of most liquids used in volumetric standards is concave with the lowest point in the center used to determine the reading. The meniscus formed by a non-wetting liquid, such as mercury (Hg), is convex with the highest point in the center. The highest point of such a meniscus is used to make the reading. The reading of a mercury barometer is a classical example of this kind. In making the reading, the observer's eye should be normal to and in the same horizontal plane as the meniscus. The illumination is adjusted to get a sharp definition of the meniscus. Elimination of parallax error is very important and can be judged by slight fluctuations of eye level that do not affect the reading.

The curvature of a meniscus is related to the surface tension of the liquid and inversely related to the diameter of the tubing in which it is formed. When reading any meniscus, it is important to ascertain that it is in an equilibrium position. Tapping of sight glasses and/or small motions of containers may be used to induce slight displacements of the meniscus. Return to the same reading is evidence of a stable meniscus.

Option A –

Option A is suitable when graduation lines extend more than 75 percent of the circumference of the sight gage area, for example with graduated neck type flasks or single mark flasks.

The position of the lowest point of the meniscus with reference to the graduation line is such that it is in the plane of the middle of the graduation line. This position of the meniscus is obtained by making the setting in the center of the ellipse formed by the graduation line on the front and the back of the tube as observed by having the eye slightly below the plane of the graduation line. This is illustrated below. The setting is accurate if, as the eye is raised and the ellipse narrows, the lowest point of the meniscus remains midway between the front and rear portions of the graduation line. By this method it is possible to observe the approach of the meniscus from either above or below the line to its proper setting.

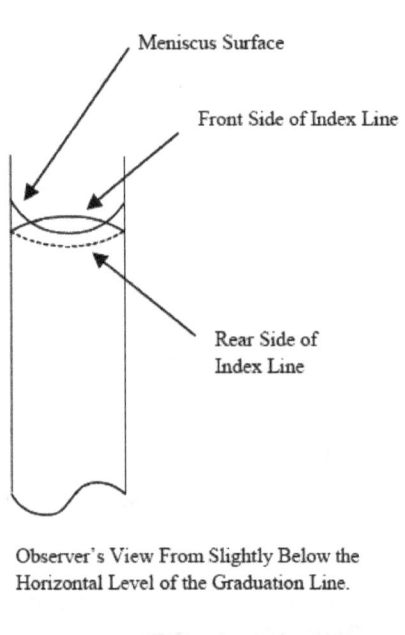

Observer's View From Slightly Below the Horizontal Level of the Graduation Line.

FRONT VIEW

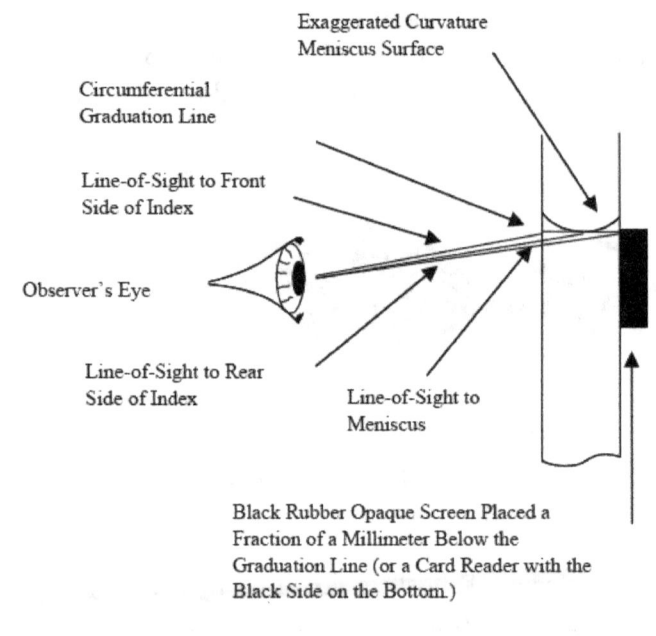

SIDE VIEW

Option B

Option B, is typically used with opaque liquids or when the graduation mark does not extend around the circumference of the volumetric standard.

The position of the lowest point of the meniscus with reference to the graduation line is horizontally tangent to the plane of the upper edge of the graduation line. The position of the meniscus is obtained by having the eye in the same plane of the upper edge of the graduation line as shown below. Offsets from reading in the same plane will produce parallax errors.

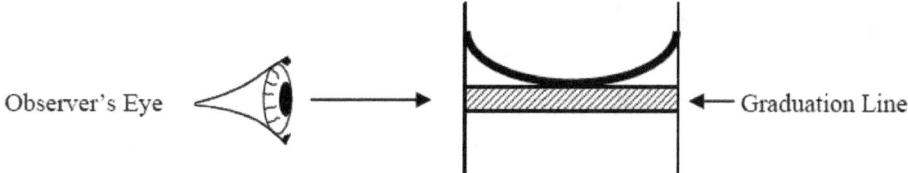

Observer's Eye ← ... → ← Graduation Line

Comparison of Option A and Option B Methods

For most practical applications the difference between these two methods is insignificant compared to the tolerances of the volumetric standards. However, a component for measurement uncertainty should be included as appropriate. When performing calibrations, using the glassware as precision standards with clear liquids, or when comparing results among laboratories the difference in meniscus setting is directly related to the visible thickness of the meniscus and the volume of liquid contained in the neck between the top of a graduation line and the middle of the graduation line as shown in the following diagram as 1) which results with Option B when clear liquids are used and 2) which results with Option A and is only possible when clear liquids are used.

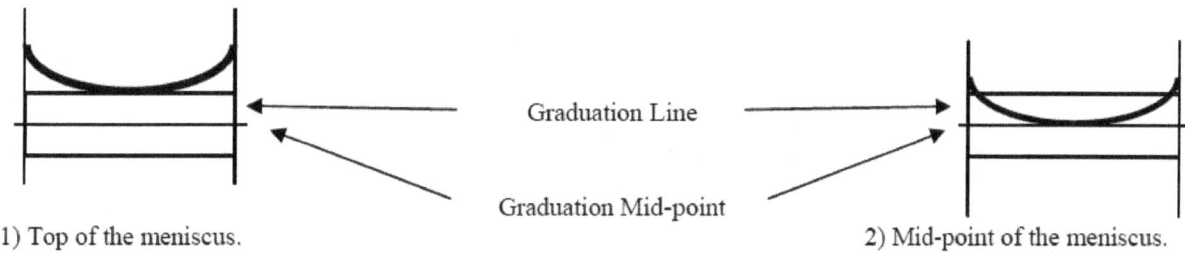

Graduation Line

Graduation Mid-point

1) Top of the meniscus.   2) Mid-point of the meniscus.

When opaque liquids are used, an additional error or correction factor can be estimated based on the volume contained between the upper and lower edges of the meniscus. The curve of the meniscus and impact of this error is dependent on the diameter of the meniscus and surface tension of the liquid being used.

Where possible, the method of reading the meniscus during calibrations should be performed in the same way in which the meniscus will be read during normal use.

Estimate potential errors in reading the meniscus, or errors in single-mark flasks by using the equation for the volume of a cylinder:

$Vol = \pi\, r^2 h$ where r is the radius of the internal diameter and h is the height of the line or volume in question. Units will be determined based on the dimensional units that are used. For example radius in cm^2 and height in cm, will provide results in cm^3 or mL. The meniscus itself often has an observed "thickness" that is larger than the graduation lines so should be considered as well.

Example:

The internal diameter of a flask is 1.5 cm (15.00 mm).
The height of the line is 0.50 mm.

$Vol = \pi\, (\dfrac{15}{2})^2\, 0.5 = 88.36$ mm^3 (0.088 mL). These values are rounded for illustrative purposes only; significance must be evaluated based on measuring instruments and flask resolution.

Experimental data obtained by reading a meniscus among multiple laboratory staff members may also be used to incorporate an estimate of uncertainty in meniscus reading.

GMP 6

Good Measurement Practice
for
Cleaning Metal Volumetric Measures

All volumetric test measures, including laboratory standards and those submitted for calibration must be clean at the time of measurement. Drainage is adversely affected by greasy and oily surfaces. Corroded surfaces raise questions that impair future use, so corrosion-damaged containers may not be worth testing.

Environmental requirements are becoming increasingly important. Many laboratories will not accept test measures and provers for calibration unless they have been suitably cleaned to remove contents such as petroleum or other chemical residues. In those cases, all surfaces (inside and out), drain lines, pumps, and hoses that may be used during the calibration process must be cleaned so that no petroleum or other product residue remains that would otherwise spill onto the laboratory floor or drain into the sewer/waste-water system.

Metal measuring vessels are best cleaned by using a non-foaming dishwashing detergent. Warm water is most effective both for cleaning and rinsing, which must be thorough. If warm water cleaning and rinsing is used, the vessel must be equilibrated to calibration temperatures before use.

The cleaning operation involves the use of a liberal amount of the detergent solution and vigorous shaking, swirling, or other motion so that the solution contacts the entire inner surface of the vessel. Depending on the condition of the surface, this operation should be repeated one or more times to ensure removal of oily films and residues. As much as possible of the detergent solution should be drained prior to the water rinses to facilitate complete removal of the detergent from the prover. Ordinarily, three rinses will be sufficient, but this should be confirmed by testing the final rinse for any visual evidence of detergent or other residues.

When detergents are not sufficient to remove oily or greasy deposits, solvents may be used, provided it is ascertained that they do not affect any coatings on the container. The surfaces should be dry before any calibrations are attempted.

If lime or scale build-up has occurred (usually only in standards that are primarily used with water), they may be cleaned with a suitable lime or scale remover. Many of these chemicals contain acids and may react with metals, so appropriate care should be taken and these chemicals should not be allowed to soak for extended periods of time. Many lime and scale removers have safety and handling requirements and may have special disposal requirements. The material safety data sheet (MSDS) should be reviewed prior to using or disposing of scale and lime removers.

After cleaning or use, vessels should be drained completely and stored in a dry place. They should be covered to prevent entrance of dust and foreign matter. Metal containers further should be protected from damage by denting and/or dropping. Once such a vessel has been visibly dented, it is difficult to ascertain whether additional changes have occurred. Even when dents have been removed and the vessel has been calibrated, visual deformities can compromise future use. Accordingly, protection from damage while stored and when in use should be a major concern of the metrologist or owner of the device.

GMP 7

**Good Measurement Practice
for
Cleaning Precision Glassware**

The volume contained in or delivered from precision glassware depends on the cleanliness of the container. Glassware must be sufficiently clean to ensure uniform wetting of the entire internal surface of the standard. When clean, the walls will be uniformly wetted and the water will adhere to the surface in a continuous film. If films of dirt or grease are present, the meniscus may crinkle at the edges and liquids will not drain properly and will leave water drops on the internal walls. Lack of cleanliness can cause errors in setting a meniscus and incomplete wetting/drainage of the calibration liquid (generally water.)

Glassware that is submitted for calibration should be cleaned (and autoclaved if needed) prior to submission to the laboratory to ensure removal of all chemical, biological, radiological, or other contaminants.

If glassware is cleaned by the laboratory, appropriate inquiries should be made to determine prior and subsequent uses along with implications of certain cleaning methods. Some methods of cleaning glassware prior to calibration or use are ineffective or detrimental and may leave chemical contamination deposits that can be detected in some chemical analysis operations.

There are a number of suitable detergents (both liquid and powder), available from laboratory supply houses that do not contain phosphates. The catalog descriptions usually indicate whether or not they contain phosphates. Laboratory detergents that contain phosphates can leave a deposit on the glass that will cause water to "bead" on the surface making it appear to be dirty, and making it difficult to properly set a meniscus. Once a phosphate deposit has occurred, it is very difficult to remove. It may be removed with hot (approx. 65 °C) sodium dichromate-sulfuric acid cleaning solution. However, use of this hot solution is hazardous and is not recommended unless proper safety equipment is available (see the next section).

If acetone is used to remove oil or grease film, always follow with alcohol (ethanol, not methanol) before drying or rinsing with water. Acetone, if allowed to dry, also leaves a film deposit.

The above solvents need not be used if the glass does not have an oil or grease film. Mechanical shaking of water and suitable non-phosphate detergents is usually adequate for cleaning glassware.

These cleaning agents do not preclude the use of other suitable methods, of which there are several. Some have special applications that would not normally be encountered by metrologists or field inspectors.

See GLP 13, Good Laboratory Practice for Drying Containers for appropriate drying techniques.

It is not necessary to dry any container marked "to deliver."

Cleaning Methods [1]

Safety Note: Material safety data sheets (MSDS) should be available, studied, and carefully followed before using any of the following cleaning methods. Chromic acid solutions are not recommended for routine use because it is a hazardous waste and hazardous to health! Suitable education or training in the handling of chemicals is highly recommended.

Frequently it is desirable to give glassware a preliminary rinse or soak with an organic solvent such as xylene or acetone to remove grease, followed by a water rinse. The rinsing with water must be done thoroughly if acid will be used later to clean the glassware.

Unless autoclaving is necessary, glassware should be cleaned as soon as possible after use to avoid setting and caking of residues. Pipets, for example, may be placed in a jar containing a weak antiseptic solution immediately after use. Autoclaving is necessary to disinfect glassware that may have been used to contain potentially dangerous biological fluids.

A selection of general cleaning methods follows:

1.   Fill with sulphuric acid-dichromate mixture and let stand. After removal of the mixture, rinse with distilled water at least six times. To make the cleaning mixture, dissolve (60 to 65) g of sodium- or potassium-dichromate by heating in (30 to 35) mL of water, cool and slowly add concentrated sulphuric acid to make one liter of solution. This solution is available from laboratory supply companies.

Note: Extreme care should be exercised in handling acidic solutions.

2.   Scrub with a 1 % to 2 % hot solution of a detergent. Rinse well after brushing. A number of suitable, commercial washing compounds are available.

3.   Fuming sulphuric acid (very hazardous material) is an excellent cleaning agent. Usually, cleaning can be accomplished by use of a comparatively small amount of acid, manipulating the vessel so that the acid comes in contact with all of the surface, and immediately emptying and rinsing.

When a piece of glassware is badly contaminated with stopcock grease (except silicone grease), it may be necessary to rinse with acetone once or twice before using one of the above methods. For silicone grease, the acetone can be omitted and the piece soaked for 30 min in fuming sulphuric acid. Warm decahydronaphthalene (decalin) also has been suggested as a solvent for silicone grease. In this case, let the piece soak for 2 h, drain, and rinse once or twice with acetone, followed by a water rinse.

------

[1] While the metrologist is not ordinarily faced with the problems for which these procedures are necessary, they are given here in the event that they are needed.

Cautions

Care should be exercised when using most cleaning solutions because they can cause skin irritations or severe burns on contact. Dilute solutions become concentrated as the water evaporates; therefore, always flush the exposed area immediately with large quantities of water.

Suitable chemical-specific-resistant goggles or a face mask should be worn to protect the eyes from splashes and rubber gloves should be worn to protect the hands. It is advisable to wear an acid resistant laboratory coat or a rubber apron to protect clothing when using strong acids for cleaning. The glassware should be handled gently to avoid breakage and also to prevent spilling acids and other cleaning fluids. All cleaning should be done in a laboratory sink or on an acid-proof laboratory bench, preferably within a fume hood, to the extent possible.

Some of the cleaning materials mentioned leave minute traces or residues unless the rinsing process is carried out thoroughly. While such traces may not be harmful if the purpose of cleaning is to prepare the glassware for calibration, they can give trouble when the glassware is used in certain laboratory operations. For example, manganese and chromium compounds, even in extreme dilution, may retard or inhibit growth of micro-organisms, and traces of phosphorus may interfere with delicate tests for this element. When glassware is to be calibrated, final rinsing must be with pure distilled or deionized water.

If an article is to be dried after cleaning, as is necessary for all vessels marked "To Contain", acetone, followed by ethyl alcohol (American Chemical Society Specification) may be used. Drying may be hastened by blowing clean, dry air into the vessel (or sucking the air through the vessel). (See GLP 13.)

Efficient air filters must be provided to remove any particles of oil or dirt from compressed air used for drying purposes.

Standard Operating Procedures

SOP No. 14

Recommended Standard Operating Procedure
for
Gravimetric Calibration of Volumetric Ware Using an Electronic Balance

1   Introduction

   1.1   Purpose of Test

          This procedure is a precision mass calibration converting mass values to volumetric values using pure water as a standard reference material. The results provide calibration of either the "to deliver" or "to contain" volume of measuring containers that may be used as volumetric measuring standards. The procedure uses gravimetric calibration principles to minimize calibration uncertainties. Accordingly, the procedure is especially useful for high accuracy calibrations. The procedure references measurement control standards to ensure the validity of the standards and the measurement process; however, additional precautions must be taken. The procedure makes use of an electronic balance and is suitable for all sizes of gravimetric calibrations only limited by the capacity and resolution of the balance. This procedure calculates the average volume based on two runs. NOTE: NIST calibrations generally make use of the average of five replicates.

   1.2   Prerequisites

      1.2.1   Verify that valid calibration certificates are available for all of the standards used in the calibration.

      1.2.2   Verify that the mass standards to be used have sufficiently small standard uncertainties for the level of calibration. Reference standards should not be used for gravimetric calibration. Weights of ASTM Class 2 or 3 or OIML Class F_1 or F_2 are needed for this procedure.

      1.2.3   Verify that the balance used is in good operating condition with sufficiently small resolution and process standard deviation, as verified by a valid control chart or preliminary experiments, to ascertain its performance quality when a new balance is put into service. The accuracy of the balance and weighing procedures should be evaluated to minimize potential bias in the measurement process. NOTE: standard deviations obtained from mass calibrations will generally not reflect the process repeatability of this procedure.

      1.2.4   Verify that the operator is experienced in precision weighing techniques and has had specific training in SOP 2, SOP 4, SOP 29, GMP 3, GMP 10, and gravimetric calibrations.

      1.2.5   Verify that an adequate supply of distilled or deionized water (see GLP 10) is available.

      1.2.6   Verify that the laboratory facilities meet the following minimum conditions to enable meeting the expected uncertainty that is achievable with this procedure:

Table 1.   **Laboratory environmental conditions.**

Procedure	Temperature	Relative Humidity
Gravimetric	20 °C to 23 °C, maxiumum change of 1 °C/h during the calibration.	40 % to 60 % ± 10 % stability / 4 h

2   Methodology

2.1   Scope, Precision, Accuracy

The procedure is applicable for the calibration of any size of measuring container that, when filled with water, will not overload the electronic balance used. Typical containers range in capacity from 1 mL to 20 L; however, this procedure is also applicable for larger provers provided that facility, equipment and standards meet the requirements in this SOP. When larger provers (e.g., 100 gal or more) are tested, also see the Test Notes in the Appendix. The precision of calibration will depend on the care exercised in adjusting the various volumes and strict adherence to the various steps of the procedure. The accuracy attainable will depend on the uncertainties of the standard weights and the air buoyancy and thermal expansion corrections that are made.

2.2   Summary

The electronic balance used is first calibrated by weighing a standard mass. The volumetric vessel to be calibrated is then weighed dry or "wetted down," depending on whether the calibration is to be made on a "To Contain" or "To Deliver" basis. The container is filled with pure water of known temperature and re-weighed. The difference in mass is used to calculate the capacity of the container at various neck graduations. The processes of this section and section 3 should be repeated as required to verify all neck graduations for which a calibrated volume is desired. Transfer vessels may be used for all procedures except for flasks and containers marked "To Contain" (TC).

2.3   Equipment and Standards

2.3.1   Electronic balance having sufficient capacity to weigh the loaded vessel. The sensitivity of the balance will be a limiting factor in the accuracy of the measurement. The resolution and repeatability must be smaller than the accepted uncertainty of the calibration. NOTE: standard deviations obtained from mass calibrations will generally not reflect the process repeatability of this procedure, therefore repeatability must be assessed using this procedure.

2.3.2   Calibrated mass standards, with recent calibration values and which have demonstrated metrological traceability to the international system of units (SI), which may be to the SI through a National Metrology Institute such as NIST.. Ordinarily, standards of ASTM Class 2 or 3 or OIML Class F_1 or F_2 weight specifications are required.

2.3.3   Calibrated thermometers, accurate to ± 0.1 °C, with recent calibration values and which have demonstrated metrological traceability to the international system of units (SI), which may be to the SI through a

National Metrology Institute such as NIST to determine water temperature.

2.3.4   Calibrated thermometer accurate to ± 0.50 °C with recent calibration values which have demonstrated metrological traceability to the international system of units (SI), which may be to the SI through a National Metrology Institute such as NIST to determine air temperature. to determine air temperature.[1]

2.3.5   Calibrated barometer accurate to ± 135 Pa (1 mm Hg) with recent calibration values which have demonstrated metrological traceability to the international system of units (SI), which may be to the SI through a National Metrology Institute such as NIST to determine air pressure. [1]

2.3.6   Calibrated hygrometer accurate to ± 10 percent with recent calibration values which have demonstrated metrological traceability to the international system of units (SI), which may be to the SI through a National Metrology Institute such as NIST to determine relative humidity.[1]

2.3.7   Distilled or deionized water (See GLP 10) of sufficient quality and quantity for the calibration. Do not use tap water for this procedure!

2.3.8   Stopwatch or other suitable timing device (does not need to be calibrated.)

2.4   General Considerations

2.4.1   Cleanliness checks

Verify that all containers to be calibrated are clean as evidenced by uniform drainage of water. No water droplets should remain on any interior surface as the water drains from the container. A reproducible "wet-down" weight is evidence for cleanliness in cases where it is not possible to visually check for uniform drainage. Use GMP 6 or 7 to clean vessels as necessary. All glassware must be meticulously cleaned, prior to calibration. When clean, the walls will be uniformly wetted. Instructions for cleaning are given in GMP 6 or 7. An exception is plastic ware, which will not be wetted. Follow manufacturer's instructions for cleaning such vessels. Do not use cleaning agents that will attack, discolor, or swell plastic ware.

2.4.2   Use water that is temperature-equilibrated with the laboratory environment. Equilibration can be achieved by storing the water in clean containers in the laboratory.

---

[1] Values from the thermometer, barometer and hygrometer are used to calculate the air density at the time of the measurement. The air density is used to make an air buoyancy correction. The accuracies specified are recommended for high precision calibration. Less accurate equipment can be used with only a small degradation in the overall accuracy of the measurement.

2.4.3 Volumetric calibrations to a marked interval are critically dependent on the setting of a meniscus. See GMP 3 for guidance in reading a meniscus.

2.4.4 Use GLP 13 as the procedure to dry any container to be calibrated on a "To Contain" basis.

2.4.5 Wet down (not used for any container calibrated "To Contain")

For glassware and hand-held test measures: Fill the container to capacity with distilled or deionized water, then empty over a 30 s period while avoiding splashing. Drain for 10 s unless another drain time is specified. (This is commonly called a "30 s pour, 10 s drain" emptying procedure.) A 30 s (± 5 s) pour followed by a 10 s drain, with the measure held between a 10 degree and 15 degree angle from vertical, is required during calibration and use for glass flasks. A wet-down is not required if a transfer vessel is used to weigh a delivered volume of water.

For provers: Fill the container to capacity with distilled or deionized water, then empty. Time the drain once the main cessation of flow is complete for 30 s and close the valve.

2.5 Calibration Procedure for Burets

2.5.1 Weigh an empty transfer vessel or flask.

2.5.2 Clamp the buret vertically on a support stand. Also clamp a plain glass test tube, large enough to hold a thermometer, in the vicinity of the buret.

2.5.3 Fill buret with water and test for absence of leaks from the tip and stopcock. Drain and fill several times to condition the buret. Fill when ready to test.

2.5.4 Drain and record delivery time, defined as the time of unrestricted flow from the zero mark to the lowest graduation mark with the stopcock fully open.

2.5.5 Fill the buret slightly above zero mark with temperature-equilibrated water and also fill the test tube that holds the thermometer. Record water temperature.

2.5.6 Set the meniscus on the zero mark and touch the tip with the wetted wall of a beaker to remove any excess water. The buret tip must be full.

2.5.7 Fully open the stopcock and discharge contents of buret into the previously weighed flask or transfer vessel. The tip should be in contact with the wall of the flask. When the level in the buret is within a few millimeters above the line being calibrated, slow the discharge and make an accurate setting. When the setting is completed, move the flask horizontally to break contact with the tip. Recheck the setting.

2.5.8 Stopper (or cover) and weigh the filled transfer vessel or flask.

2.5.9 Measure and record the temperature of water in the container.

2.5.10 Test the next interval in the same manner - from the zero mark to the next interval of test.

2.5.11 For burets with a specified waiting time, empty as in 2.5.6 to within a few millimeters of the calibration mark. Pause for the specified waiting time (e.g., 10 s), then adjust the meniscus to the graduation line as in 2.5.6.

2.5.12 Make a duplicate determination for each interval (Run 2).

2.5.13 Calculate the volume for each interval as described in Section 3.

2.6 Calibration Procedure for Pipets (One-Mark)

2.6.1 Weigh an empty transfer vessel or flask.

2.6.2 Fill the pipet to the index mark and measure the delivery time with the tip in contact with the internal surface of a beaker.

2.6.3 Refill the pipet by suction, slightly above the index line. Record the water temperature. Wipe tip with filter paper, then slowly lower level to the index line, using a stopcock or pinch clamp for fine control. The tip must be in contact with the wetted wall of a beaker while this setting is being made. Do not remove any water remaining on tip at this time.

2.6.4 Hold pipet in a vertical position and deliver water into a previously weighed weighing flask, with the tip in contact with the inside wall or neck.

2.6.5 After flow has ceased, wait two seconds then remove the pipet from contact with the flask.

2.6.6 Stopper (or cover) and weigh the filled transfer vessel or flask.

2.6.7 Make a duplicate determination for each interval (Run 2).

2.6.8 Calculate the volume as described in Section 3.

2.7 Calibration of Flasks (To Contain) – Transfer vessel may not be used.

2.7.1 Clean and dry flask as described in GLP 13, then stopper and weigh the flask. Record the Empty Flask value.

2.7.2 Place an appropriate sized funnel in neck and fill flask to just below the reference graduation while maneuvering the flask to wet the entire neck below the stopper.

2.7.3 Let stand for two minutes then adjust the meniscus to the reference graduation line.

2.7.4 Determine the temperature of the water by putting some in a beaker or test tube containing a thermometer.

2.7.5 Weigh the filled flask and record the Filled Flask mass value.

2.7.6   Make a duplicate determination (Run 2) after drying the flask per GLP 13.

2.7.7   Perform volume calculations as described in Section 3.

2.8   Calibration of Flasks (To Deliver)

2.8.1   Clean but do not dry a transfer vessel or flask.

2.8.2   Weigh the empty transfer vessel or flask and record the mass.

2.8.3   Place an appropriately sized funnel in the neck and fill the flask to just below the reference graduation while maneuvering the flask to wet the entire neck below the stopper.

2.8.4   Let the flask stand for two minutes, then adjust the meniscus to the reference graduation line, then weigh the full vessel, with cap or stopper.

2.8.5   Empty the flask over a 30-second period by gradually inclining it so as to avoid splashing. When the main drainage has ceased, hold the flask in vertical position for 10 seconds unless another drain time is specified, then touch off the drop of water adhering to the top of the flask.

2.8.6   Cover the flask or transfer vessel and reweigh.

2.8.7   Make a duplicate determination (Run 2).

2.8.8   Calculate the volume of the flask as described in Section 3.

2.9   Calibration of Other Volumetric Glassware.

2.9.1   Measuring Pipets - Calibrate in a manner similar to that used to calibrate burets.

2.9.2   Graduated Cylinders - Calibrate in a manner similar to that used for flasks.

2.10   Calibration of Slicker-Plate Type Standards

2.10.1   Condition the slicker-plate type standards with several wet-down runs to fully ensure wet down and smooth valve operation.

2.10.2   Weigh an empty transfer vessel and record the mass.

2.10.3   Fill the slicker-plate standard to just above the rim of the standard. Record the water temperature. Slide the slicker plate across the level top. Set the transfer vessel below the nozzle to ensure all volume is transferred into the vessel and that no splashing occurs.

2.10.4   Open the slicker-plate standard valve and remove the plate simultaneously and smoothly to deliver the water into the transfer vessel. Time the drain for 30 s after the cessation of the main flow and close the valve. Cover the transfer vessel and move it from beneath the standard to ensure additional drops of water are not transferred.

2.10.5 Weigh the filled transfer vessel and record the mass.

2.10.6 Make a duplicate determination for each interval (Run 2).

2.10.7 Calculate the volume as described in Section 3.

2.11 Calibration of Graduated Neck Type Provers

2.11.1 Condition the graduated neck type standards with several wet-down runs to fully ensure wet down and smooth valve operation.

2.11.2 Weigh an empty transfer vessel and record the mass.

2.11.3 Fill the standard to the nominal graduation mark. Record the water temperature. Set the transfer vessel below the nozzle to ensure all volume is transferred into the vessel and that no splashing occurs.

2.11.4 Open the standard valve smoothly to deliver the water into the transfer vessel. Time the drain for 30 s after the cessation of the main flow and close the valve. Cover the transfer vessel and move it from beneath the standard to ensure additional drops of water are not transferred.

2.11.5 Weigh the filled transfer vessel and record the mass.

2.11.6 Make a duplicate determination for each interval (Run 2).

2.11.7 Calculate the volume as described in Section 3.

2.12 Weighing Procedures

2.12.1 Weighing (Option A – One point balance calibration)

2.12.1.1 Zero the balance and record reading as O_1. Place a standard mass, M_S, on the balance platform (where possible, M_S should be slightly larger than the mass of the filled vessel.) Record reading as O_2.

2.12.1.2 Zero the balance. Place dry or "wet-down" container or empty transfer vessel on the balance platform, as appropriate, and record reading as O_3.[2] Caution: all containers must be dry on the outside for all weighing.

2.12.1.3 Fill container to its reference mark. Read and record the temperature of the water used to fill the container. Carefully adjusting the meniscus (if present) to minimize filling error (see GMP No. 3). Zero the balance. Weigh the filled vessel or transfer vessel and record the reading as O_4.

2.12.1.4 Immediately after weighing, check the temperature of the water in the filled container. If the temperature differs by more than 0.2 °C from that of 2.4.4.3, refill and reweigh as described in 2.4.4.3.

2.12.1.5 Record air temperature, barometric pressure, and relative humidity at the time of the above measurements.

2.12.1.6 Make a duplicate determination (Run 2).

---

[2] When calibrating "to deliver" vessels, O_4 may be measured before O_3. If a transfer vessel is used, the drained mass or empty mass is usually measured before the filled mass.

2.12.2 Weighing (Option B – Two point balance calibration.)

    2.12.2.1   Zero the balance and record reading as O_1. Place a standard mass, M_{S1}, on the balance platform (where possible M_{S1} should be slightly larger than the mass of the drained vessel, dry vessel, or empty transfer vessel.) Record reading as O_2.

    2.12.2.2   Zero the balance. Place dry or "wet-down" container on balance platform, as appropriate, and record reading as O_3. Caution: all containers must be dry on the outside for all weighing.

    2.12.2.3   After removing empty vessel, zero the balance and record reading as O_4. Place a standard mass, M_{S2}, on the balance platform (where possible M_{S2} should be slightly larger than the mass of the filled vessel.) Record reading as O_5.

    2.12.2.4   Fill standard container to its reference mark. Read and record the temperature of the water used to fill the container. Carefully adjusting the meniscus (if present) to minimize filling error (see GMP 3). Zero the balance. Weigh the filled vessel or transfer vessel and record reading as O_6.

    2.12.2.5   Immediately after weighing, check the temperature of the water in the filled container. If the temperature differs by more than 0.2 °C from that of 2.4.5.4, refill and reweigh as described in 2.4.5.4.

    2.12.2.6   Record air temperature, barometric pressure, and relative humidity at the time of the above measurements.

    2.12.2.7   Make a duplicate determination (Run 2).

3   Calculations

    3.1   Option A – One-point balance calibration. Compute the volume, V_t, for each determination using the equation:

$$V_t = \left(O_4 - O_3 \right) \left(\frac{M_s}{O_2 - O_1} \right) \left(1 - \frac{\rho_a}{\rho_s} \right) \left(\frac{1}{\rho_w - \rho_a} \right) \qquad \text{Eqn. 3.1}$$

3.2 Option B – Two point balance calibration. Compute the volume, V_t, for each determination using the equation:

$$V_t = \left[O_{6(filled)} \frac{M_{s2}\left(1 - \dfrac{\rho_a}{\rho_s}\right)}{(O_5 - O_4)} - O_{3(drained)} \frac{M_{s1}\left(1 - \dfrac{\rho_a}{\rho_s}\right)}{(O_2 - O_1)} \right] \left(\frac{1}{\rho_w - \rho_a} \right) \quad \text{Eqn. 3.2}$$

Table 2. **Variables for volume equations.**

Variable	Description
M_S, M_{s1}, M_{s2}	mass of standards (i.e., true mass, vacuum mass) (g)
ρ_s	density of M_S (g/cm³)
ρ_w	density of water at the temperature of measurement (g/cm³)
ρ_a	density of air at the conditions of calibration (g/cm³)
V_t	represents either the "to contain" or "to deliver" volume (depending on whether O_3 or O_6 represent a dry or a "wet down" container at the temperature of the measurement) (cm³ or mL)

3.3 Glassware is typically calibrated to 20 °C. Compute V_{20}, the volume at 20 °C, for each run, using the expression:

$$V_{20} = V_t \left[1 - \alpha \left(t - 20 \right) \right]$$

where α is the cubical coefficient of expansion of the container being calibrated, (see NISTIR 6969, Table 9.10), and, t, is the temperature (°C) of the water. Compute the mean V_{20} for the duplicate measurements.

3.4 If using a different reference temperature, use the following equation and take care to match the cubical coefficient of expansion units with the units of temperature:

$$V_{ref} = V_t \left[1 - \alpha \left(t - t_{ref} \right) \right]$$

3.5. Other Reference temperatures may be used. Common reference temperatures for other liquids follow:

Commodity	Reference Temperature
Frozen food labeled by volume (e.g., fruit juice)	-18 °C (0 °F)
Beer	3.9 °C (39.1 °F)
Food that must be kept refrigerated (e.g., milk)	4.4 °C (40 °F)
Distilled spirits or petroleum	15.56 °C (60 °F)
Petroleum (International Reference)	15 °C (59 °F)
Wine	20 °C (68 °F)
Unrefrigerated liquids (e.g., sold unchilled, like soft drinks)	20 °C (68 °F)
Petroleum (Hawaii)	26.67 °C (80 °F)

4   Measurement Assurance

    4.1   Duplicate the process with a suitable check standard (See GLP 1, SOP 30, and NISTIR 6969, Sec. 7.4) or conduct replicate measurements per SOP 17 or 20. Average values of the range or standard deviation of similarly sized flasks or volumetric standards may be tracked on a single range chart or standard deviation chart according to SOP 17 or 20. A minimum of 12 replicate measurements are required to establish initial process limits.

    4.2   Plot the check standard volume and verify it is within established limits. A t-test may be incorporated to check the observed value against an accepted value.

    4.3   The mean of the check standard is used to evaluate bias and drift over time and may be used to identify or signify problems with the volume standard or changes in water quality.

    4.4   Check standard observations are used to calculate the standard deviation of the measurement process, s_p.

5 Assignment of Uncertainties

The limits of expanded uncertainty, U, include estimates of the standard uncertainty of the mass standards used, u_c, plus the uncertainty of measurement, s_p, at the 95 percent level of confidence. See SOP 29 for the complete standard operating procedure for calculating the uncertainty.

    5.1   The standard uncertainty for the standards, u_s, is obtained from the calibration report. The combined standard uncertainty, u_c, is used and not the expanded uncertainty, U, therefore the reported uncertainty for the standard will usually need to be divided by the coverage factor k. Multiple mass standards are often used, so see SOP 29 for treatment of dependencies.

    5.2   Standard deviation of the measurement process from control chart performance (See SOP No. 17 or 20.)

The value for s_p is obtained from the control chart data for check standards when a check standard is available. It may also be estimated based on replicate measurements over time. Replicate measurements over time may provide a pooled standard deviation that may be used or the average range is used to estimate the standard deviation per SOP 17 or 20. This value incorporates a repeatability factor related to the precision of the weighings and the setting of the meniscus when present, but does not include uncertainties associated with errors in reading the meniscus.

5.3   Include uncertainties associated with the reading of the meniscus when present. See GMP 3 for details.

5.4   Other standard uncertainties usually included at this calibration level include uncertainties associated with water temperature measurements, thermometer accuracy, calculation of air density, standard uncertainties associated with the density of the standards used, coefficients of expansion, viscosity or surface effects on the volume of liquid clinging to vessel walls after draining, improper observance of drainage times, and the lack of internal cleanliness.

Additional References:

Bean, V. E., Espina, P. I., Wright, J. D., Houser, J. F., Sheckels, S. D., and Johnson, A. N., NIST Calibration Services for Liquid Volume, NIST Special Publication 250-72, National Institute of Standards and Technology, Gaithersburg, MD, (2006)

Appendix A
Gravimetric Calibration Data Sheet (Option A)

Laboratory data and conditions:

				Before	After
Vessel ID		Operator			
Material		Date			
Cubical Coefficient of Expansion					
Balance		Temperature			
Load		Pressure			
Standard deviation of the process, from control chart, s_p		Relative Humidity			
Degrees of Freedom		Water temperature (for reference)			

Mass standard(s) data:

ID (Note ID and for Filled or Empty Load)	Nominal	Mass Correction*	Expanded Unc: From cal. report	Unc: k factor	Density g/cm^3
S					
S					
S					
S					
S					

*Mass Correction = *True Mass* if using buoyancy correction. Density is required for buoyancy corrections.

Observations:

Run 1	Weights		Balance Observations, Units_____
1	Zeroed Balance	O_1	0.00
2	M_S	O_2	
3	Empty or Drained	O_3	
4	Filled	O_4	
	t_w:		

Run 2	Weights		Balance Observations, Units_____
1	Zeroed Balance	O_1	0.00
2	M_S	O_2	
3	Empty or Drained	O_3	
4	Filled	O_4	
	t_w:		

Example
Gravimetric Calibration Data Sheet (Option A)

Laboratory data and conditions:

				Before	After
Vessel ID	321	Operator		10/1/99	
Material	Soda-lime glass	Date		GH	
Cubical Coefficient of Expansion	0.000025 / °C			Before	After
Balance	LC 5100	Temperature		22.4 °C	22.6 °C
Load	2 L	Pressure		747.6 mm Hg	748.0 mm Hg
Standard deviation of the process, from control chart, s_p	0.042 mL	Relative Humidity		43 %	47 %
Degrees of Freedom	216	Water temperature (for reference)		22.8 °C	22.6 °C

Mass standard(s) data:

ID (Note ID and for Filled or Empty Load)	Nominal	Mass Correction*	Expanded Unc: From cal. report	Unc: k factor	Density g/cm³
S	2 kg	0.000123 g	2 mg	2	7.95
S					
S					
S					
S					

*Mass Correction = *True Mass* if using buoyancy correction. Density is required for buoyancy corrections.

Observations:

Run 1	Weights		Balance Observations, Units__g____
1	Zeroed Balance	O_1	0.000
2	M_S	O_2	2000.003
3	Empty or Drained	O_3	654.729
4	Filled	O_4	2648.747
	t_w:		22.8 °C
Run 2	Weights		Balance Observations, Units_____
1	Zeroed Balance	O_1	0.000
2	M_S	O_2	1999.998
3	Empty or Drained	O_3	667.351
4	Filled	O_4	2661.365
	t_w:		22.6 °C

Calculate the air density (SOP 2) ρ_a:

$$\rho_a = 1.169\,625\,\text{mg/cm}^3 \ = 0.001\,169\,625\,\text{g/cm}^3$$

Round the results to 9 digits.

Calculate (or look up) the density of the water, ρ_w:

$$22.8\ °\text{C} \ = 0.997\,586\,95\,\text{g/cm}^3$$

$$22.6\ °\text{C} \ = 0.997\,633\,78\,\text{g/cm}^3$$

Round the results to 8 digits.

Compute the volume, V_t, for each determination using the equation:

$$V_t = \left(O_4 - O_3\right)\left(\frac{M_s}{O_2 - O_1}\right)\left(1 - \frac{\rho_a}{\rho_s}\right)\left(\frac{1}{\rho_w - \rho_a}\right)$$

Run 1:

$$V_t = \left(2\,648.747 - 654.729\right)\left(\frac{2\,000.000\,123}{2\,000.003 - 0}\right)\left(1 - \frac{0.001\,169\,625}{7.95}\right)\left(\frac{1}{0.997\,586\,95 - 0.001\,169\,625}\right)$$

$$V_t = \left(1\,994.018\right)\left(0.999\,998\,562\right)\left(0.999\,852\,877\right)\left(1.003\,446\,462\right)$$

$$V_t = \left(1\,994.018\right)\left(1.003\,446\,462\right) = 2\,000.890\,307\,\text{mL}$$

Compute V_{20}, the volume at 20 °C, for Run 1 using the expression:

$$V_{20} = V_t\left[1 - \alpha\left(t - 20\right)\right]$$

$$V_{20} = 2\,000.890\,307\left[1 - 0.000\,025\left(22.8 - 20\right)\right] = 2\,000.750\,244\,\text{mL}$$

Run 2:

$$V_t = \left(2\,661.365 - 667.351\right)\left(\frac{2\,000.000\,123}{1999.998 - 0}\right)\left(1 - \frac{0.001\,169\,625}{7.95}\right)\left(\frac{1}{0.997\,633\,78 - 0.001\,169\,625}\right)$$

$$V_t = \left(1\,994.014\right)\left(1.000\,001\,062\right)\left(0.999\,852\,877\right)\left(1.003\,548\,392\right)$$

$$V_t = \left(1\,994.014\right)\left(1.003\,401\,812\right) = 2000.797\,261\,\text{mL}$$

Compute V_{20}, the volume at 20 °C, for Run 2 using the expression:

$$V_{20} = V_t \left[1 - \alpha \left(t - 20 \right) \right]$$

$$V_{20} = 2000.797\,261 \left[1 - 0.000025 \left(22.6 - 20 \right) \right] = 2000.667209 \text{ mL}$$

Calculate the mean V_{20} :

$$\overline{V}_{20} = \frac{(2\,000.750\,244 + 2\,000.667\,209)}{2} = 2\,000.708\,727 \text{ mL}$$

Calculate the uncertainty for the calibration:

$$U = u_c * 2$$

$$u_c = \sqrt{u_s^2 + s_p^2 + u_o^2}$$

The uncertainty for the standard must be divided by the k factor for the standard, u_s. All values must be represented in like units.

$$U_s = 2 \text{ mg}, k = 2 \qquad u_s = 2 \text{ mg} / 2 = 1 \text{ mg} \qquad \approx 1 \text{ mL}$$

$$s_p = 0.042 \text{ mL}$$

$$u_o = 0.00018 \text{ mL}$$

$$u_c = \sqrt{(0.001)^2 + (0.042)^2 + (0.00018)^2}$$

$$u_c = 0.042\,012 \text{ mL}$$

$$U = 0.042\,012 * 2 = 0.084\,024 \text{ mL}$$

The volume correction and uncertainty are reported as follows when rounded to two significant digits according to NISTIR 6969, GMP 9:

$$V_{20} = 2000.709 \text{ mL} \pm 0.084 \text{ mL}$$

Appendix B
Gravimetric Calibration Data Sheet (Option B)

Laboratory data and conditions:

				Before	After
Vessel ID		Operator			
Material		Date			
Cubical Coefficient of Expansion				Before	After
Balance		Temperature			
Load		Pressure			
Standard deviation of the process, from control chart, s_p		Relative Humidity			
Degrees of Freedom		Water temperature (for reference)			

Mass standard(s) data:

ID (Note ID and for Filled or Empty Load)	Nominal	Mass Correction*	Expanded Unc: From cal. report	Unc: k factor	Density g/cm³
S					
S					
S					
S					
S					

*Mass Correction = *True Mass* if using buoyancy correction. Density is required for buoyancy corrections.

Observations:

Run 1	Weights		Balance Observations, Units_____
1	Zeroed Balance	O_1	0.000
2	M_{s1}	O_2	
3	Empty or Drained	O_3	
4	Zeroed Balance	O_4	0.000
5	M_{S2}	O_5	
6	Filled	O_6	
	t_w:		

Run 2	Weights		Balance Observations, Units_____
1	Zeroed Balance	O_1	0.000
2	Zeroed Balance	O_2	
3	M_{s1}	O_3	
4	Empty or Drained	O_4	0.000
5	Zeroed Balance	O_5	
6	M_{S2}	O_6	
	t_w:		

Appendix

Test Notes for Large Provers

1   Pour and drain times.

It is impractical to completely drain a filled container, because some of the contents will remain as a film. By strict adherence to a specified procedure, the residual contents can be held essentially constant so that, reproducible calibration constants can be obtained. The conditions conventionally selected are as follows:

    a   For bottom-drain containers: open drain valve fully and allow contents to discharge at maximum rate. When flow ceases, wait 30 s, close valve, and touch off any drops adhering to spout.

    b   For pour-type containers: pour contents by gradually tilting container to an 85 ° angle, so that virtually all is delivered in 30 s. Allow to drain for an additional 10 s, then touch off any drops adhering to the lip.

The instructions described above must be precisely followed during calibration and use of the calibrated vessels.

2   Evaporation losses.

A suitable cap should be placed on the top of open vessels to minimize evaporation losses. If used, the cap must be included in all weighings.

3   Slicker-plate.

When a slicker-plate standard is calibrated, the plate should be used to fix the water level in it. This plate must be weighed with the standard during each such operation (unless a transfer vessel is used).

SOP 16

**Standard Operating Procedure for
Calibration of Measuring Flasks
Volume Transfer Method**

1   Introduction

1.1   Purpose of Test

This procedure describes a method for volume transfer calibration of the "to deliver" volume of measuring flasks from calibrated volumetric standards. The test itesm are then used as volumetric measurement standards, often for packaged commodity verification. If "to contain" calibrations are to be performed, appropriate modifications are needed in the procedure to ensure complete drying of the flask between calibration runs.

1.2   Prerequisites

1.2.1   Verify that valid calibration certificates are available for all of the standards used in the calibration.

1.2.2   Verify that the standards to be used have sufficiently small standard uncertainties for the intended level of calibration.

1.2.3   Verify that the standard burets and pipets are clean, free of leaks, and in good operational condition.

1.2.4   Verify the availability of an adequate supply of pure distilled or deionized water (See GLP 10).

1.2.5   Verify that the operator has had specific training and is proficient in SOP 16, GMP 3, GMP 7, and is familiar with the operating characteristics and conditioning of pipets and burets.

1.2.6   Verify that the laboratory facilities meet the minimum conditions shown in Table 1 to meet the expected uncertainty that is achievable with this procedure.

Table 1.   Laboratory environmental conditions.

Procedure	Temperature	Relative Humidity
Volume Transfer	18 °C to 27 °C, maximum change 1 °C/h during the calibration	40 % to 60 % ± 20 % max change / 4 h

2   Methodology

    2.1   Scope, Precision, Accuracy

The method is applicable for the calibration of any size of measuring flask for which standard pipets of comparable volume are available. Typical flasks have volumes in the range of 100 mL to 5 L (1 gill to 1 gallon). The precision of calibration depends on the care exercised in adjusting the various volumes and strict adherence to the various steps of the procedure. The accuracy will depend on the accuracy of calibration of the volumetric standards used, accuracy in reading the meniscus, together with the precision of the comparison. Clean glassware and strict adherence to the drainage instructions are essential for precise and accurate results.

    2.2   Summary

The flask to be calibrated is given an initial wet-down using the correct drain procedure, then nearly filled with water delivered from a calibrated pipet. (This procedure relies on calibrated pipets that deliver slightly less than the full nominal volume.) Additional water is added to the test flask from a calibrated buret until the meniscus in the test item coincides with the calibration graduation mark. The sum of the volumes delivered into the wetted flask is equivalent to its "to deliver" volume. The process described assumes that the flask is marked with a 10-second drain time. If the flask is marked with a drain time other than 10 s, the specified drain time should be used rather than the 10 s drain time described. A 30-second drain time should be used when a flask has no drain time specified.

    2.3   Equipment

        2.3.1   Calibrated standard pipet(s) of suitable volume with recent calibration values and which have demonstrated metrological traceability to the international system of units (SI), which may be to the SI through a National Metrology Institute such as NIST. These standards are typically made of borosilicate glass.

        2.3.2   Calibrated standard buret(s) of 10 mL (or 120 minim[1]) capacity with recent calibration values and which have demonstrated metrological traceability to the international system of units (SI), which may be to the SI through a National Metrology Institute such as NIST. These standards are typically made of borosilicate glass.

        2.3.3   Calibrated thermometers, accurate to ± 0.1 °C, with recent calibration values and which have demonstrated metrological traceability to the international system of

---

[1] 1 [US] minim = 0.061615496 milliliter (mL). 120 minim standards were provided to the States as reference standards for use in volume transfer calibrations. These standards have higher resolution than metric standards provided at the same time.

units (SI), which may be to the SI through a National Metrology Institute such as NIST to determine water temperature.

2.3.4   Distilled or deionized water (See GLP 10) of sufficient quality and quantity for the calibration. Do not use tap water for this procedure to ensure the reference standards are maintained with valid reference values!

2.3.5   Meniscus reading device (See GMP 3).

2.3.6   Stopwatch or other suitable timing device (does not need to be calibrated).

2.4   Procedure

2.4.1   Cleanliness check

Verify that all glassware, including the standards used and the vessels to be calibrated, are internally clean, as evidenced by uniform drainage of water. No water droplets should remain on the internal surfaces as the water drains from the vessels. If this occurs, the glassware must be cleaned with suitable agents and rinsed with pure water until uniform drainage is obtained. (See GMP 7).

2.4.2   Wet-down of pipet and flask

2.4.2.1 Fill the standard pipet to until water flows from the top tip, then completely drain into an empty vessel. Check for uniform drainage. Touch off the pipet outlet tip against the container wall to remove excess droplets and to establish a constant tip retention volume. The pipet and bore of the delivery side of the stopcock should appear to be "empty" and the delivery tip should contain a small volume of water retained in it. This is the "wet-down" condition of the pipet. It must be done at the beginning of each test sequence. Ensure that the outside of the standard pipet is free of water droplets that may fall into the unknown flask.

2.4.2.2 Refill the pipet and use it to fill the flask to be calibrated. Empty the flask using a gentle pour in a 30 s ± 5 s period by gradually inclining the flask so as to avoid splashing of the walls as much as possible. When the main drainage stream has ceased, the flask will be nearly vertical. Allow an additional 10 second drainage after discharge of the flask contents while holding the flask at a 10 ° to 15 ° angle from vertical, then touch off the rim of the flask to remove any drops adhering to it. At the same time, check that uniform drainage has been achieved. This establishes the "to deliver" condition of the flask. Note that some small amount of water will remain in the flask. Each flask to be calibrated must be given this "wet-down" treatment immediately before the start of the test.

2.4.3   Conditioning the buret

2.4.3.1 Fill the standard buret to overflow and drain several times to verify uniform drainage. Refill with water. Note that the stopcock bore and delivery tip should be filled with water at all times, in contrast to the condition for the pipet.

2.4.3.2 Touch off the delivery tip against the wall of the receiving vessel, to remove any droplet adhering to the external surface of the tip. This is an operation that must be done every time a measured volume of water is delivered from the buret to ensure consistent volume delivery.

2.4.4   Calibration

2.4.4.1 Run 1. Fill the standard pipet to overflow. The delivery bore and the delivery tip will be empty, except for the small volume retained in the tip. Place the inside of the flask neck in contact with the tip of the pipet or buret, to avoid splashing but in a manner that does not block the flow of water.

2.4.4.2 Deliver the contents of the pipet into the "wet-down" flask. Repeat as necessary to nearly fill the flask to the calibration line. The value recorded for the standard pipet will be the total of the deliveries used.

2.4.4.3 Add water to the flask from the buret until the meniscus coincides with the calibration mark. (See GMP 3 for instructions on how to read a meniscus.) Read the volume of water delivered from the buret and record on a suitable data sheet such as the one in the Appendix. Note: If multiple deliveries of the complete buret contents are required, ensure that at no time the buret water level drops below the lowest graduation mark, or the entire process must be started over. The value recorded for the standard buret will be the sum of the delivered volumes.

2.4.4.4 Empty the measuring flask as described in 2.4.2.2 to re-establish "wet-down."

2.4.4.5 Record all data using the form given in the Appendix or a similar format.

2.4.5 Run 2. Replicate Measurement. Repeat the procedure described in 2.4.4. The test measure must be capable of repeating to 0.01 % of the test volume during calibration. Repeatability problems may be due to contamination, poor meniscus readings, lack of cleanliness, or conduct of the calibration in an unstable environment. Repeatability problems must be corrected before calibration can be completed.

2.4.6 Calibration Note: If the test vessel has a graduated neck (in addition to the nominal mark) additional graduations should be inspected for proper marking and calibrated as well.

3 Calculations

3.1 Compute the individual "to deliver" volumes, V_{TD}, for Run 1 and Run 2.

$$V_{TD} = V_P + V_B \qquad \text{Eqn. 1}$$

Table 2.   **Variables for volume equation.**

Variable	Description
V_{TD}	"To deliver" Volume
V_P	"To deliver" calibrated volume of the standard pipet
V_B	Volume of water delivered from the buret, corrected for any calibration values

3.2 Temperature correction

For calibration of glassware that is not made of borosilicate glass, not only should the above temperature conditions be realized but the water temperature must also be known. The water used must be stored in the laboratory until its temperature is equilibrated with its surroundings and the temperature of the discharged water is measured, using the calibrated thermometer. The "To Deliver" volume of the flask, corrected to the reference temperature of 20 °C, V_{TD20}, is computed, using the expression:

$$V_{TD20} = V_{TD}\left[1 + (\alpha - \beta)(t - 20)\right] \qquad \text{Eqn. 2}$$

Table 3.   **Variables for volume temperature correction equation.**

Variable	Description
V_{TD}	observed delivered volume as computed in 3.1
V_{TD20}	"To deliver" volume at the reference temperature of 20 °C
t	Temperature of water at time of calibration
α	Cubical coefficient of expansion of glass of standard (e.g., 0.000010 °C⁻

| | ^{1} for borosilicate glass) |
| β | Cubical coefficient of expansion of glass of flask (e.g., 0.000025 °C^{-1} for soda-lime glass) |

The volumetric standards are typically made of borosilicate glass. If borosilicate flasks are calibrated, no temperature correction is required. This procedure relies on the fact that since deliveries are made over a short period of time, the temperature remains the same in the standard as in the unknown. *When standards and flasks are not of the same material, the water temperature change must be less than 0.5 °C while the water is in the standards and the flask.*

3.3 Report the average volume V_{TD20} as the value for the flask after the temperature correction has been applied to each Run.

4 Measurement Assurance

4.1 Duplicate this process with a suitable check standard (Quality Assurance Reference Flask, QARF) or have a suitable range of check standards for the laboratory with similar size graduations and neck diameters. See SOP 17 and SOP 30.

4.2 Plot the check standard volume and verify that it is within established limits. Alternatively a *t*-test may be incorporated into the process to check the observed value against an accepted reference value. (See NISTIR 6969, Section 8).

4.3 The mean of the check standard value is used to evaluate bias and drift over time.

4.4 Check standard observations are used to calculate the standard deviation of the measurement process.

5 Assignment of Uncertainties

The limits of expanded uncertainty, U, include estimates of the standard uncertainty of the laboratory volumetric standards used, u_s, plus the standard deviation of the process, s_p, at the 95 percent level of confidence. See SOP 29 for the complete standard operating procedure for calculating the uncertainty.

5.1 The standard uncertainty for the standard, u_s, is obtained from the calibration reports for the pipet and buret used for the calibration. The combined standard uncertainty, u_c, from the calibration report must be used and not the expanded uncertainty, U. Therefore, the reported uncertainty for the standard will usually need to be divided by a coverage factor, k. Multiple standards are used in this procedure; however, they are typically calibrated independently. Multiple standard uncertainties are handled according to SOP 29.

5.2 The standard deviation of the measurement process is obtained from control chart performance (See SOP 17 or 20, and SOP 30).

5.3 The value for s_p is obtained from the control chart data of the check standard using the same volume transfer procedures on flasks of comparable size and neck dimensions.

5.4   Other standard uncertainties usually included at this calibration level include uncertainties associated with the ability to read the meniscus, only part of which is included in the process variability (See GMP 3), the cubical coefficient of expansion for the flask under test, use of proper temperature corrections, the accuracy of temperature measurements taking into consideration potential gradients in the glassware during calibration, round robin data showing reproducibility, environmental variations over time, and bias or drift of the standard.

6   Report

6.1   Report results as described in SOP 1, Preparation of Calibration/Test Results, with the addition of the following items:

6.1.1   Volume, applicable reference temperature, uncertainty, reference temperature, material, thermal coefficient of expansion (assumed or measured), construction, any identifying markings, tolerances (if appropriate), laboratory temperature, water temperature at time of test, barometric pressure, relative humidity, and any out-of-tolerance conditions.

7   Alternative Reference Temperatures

7.1   Reference temperatures other than 20 °C (68 °F) may occasionally be used. Common reference temperatures for other liquids follow:

Commodity	Reference Temperature
Frozen food labeled by volume (e.g., fruit juice)	- 18 °C (0 °F)
Beer	3.9 °C (39.1 °F)
Food that must be kept refrigerated (e.g., milk)	4.4 °C (40 °F)
Distilled spirits or petroleum	15.56 °C (60 °F)
Petroleum (International Reference)	15 °C (59 °F)
Wine	20 °C (68 °F)
Unrefrigerated liquids (e.g., sold unchilled, like soft drinks)	20 °C (68 °F)
Petroleum (Hawaii)	26.67 °C (80 °F)

7.2   When alternative reference temperatures are used, at the request of the user of the field standard, and when the glassware is *not* marked with a reference temperature, the calibrated value of the standard may be reported at a different reference temperature, t_{ref}, by using this equation:

$$V_{TDtref} = V_{TD}\left[1 + (\alpha - \beta)(t - t_{ref})\right] \qquad \text{Eqn. 3}$$

Additional References:

Bean, V. E., Espina, P. I., Wright, J. D., Houser, J. F., Sheckels, S. D., and Johnson, A. N., NIST Calibration Services for Liquid Volume, NIST Special Publication 250-72, National Institute of Standards and Technology, Gaithersburg, MD, (2006).

Appendix

Calibration of Measuring Flask
(Volume Transfer Method)

Laboratory data and conditions:

Test Number		Date			
Vessel ID		Operator			
Nominal Volume				Before	After
Coefficient of Expansion		Temperature			
Process standard deviation from control chart, s_p		Pressure			
Degrees of Freedom		Relative Humidity			

Volume standard(s) data:

ID (enter interval if needed)	Nominal	Calibrated Volume	Volume Correction	Coefficient of Expansion	Uncertainty	k factor
Pipet						
Buret						
Buret						
Buret						

Observations:

	Run 1	Run 2
Calibrated Pipet Volume: V_p		
Buret Volume:		
Final Reading		
Initial Reading		
Difference		
Calibration Correction (if required)		
Corrected Buret Volume, V_B		
$V_{TD} = V_P + V_B$		
Water temperature at time of test (t)		
$V_{TD20} = V_{TD}\left[1 + (\alpha - \beta)(t - 20)\right]$		
Average V_{TD20}		

SOP 17

Standard Operating Procedure for
Control Charts of Laboratory Owned Check Standards

1   Introduction

   1.1 Purpose

   This procedure may be used to develop and maintain control charts to monitor the statistical control laboratory owned check standards.

   1.2 Prerequisites

      1.2.1 The procedure to be monitored must match the calibration procedure that is used.

      1.2.2 Either a check standard at each nominal value is used or a set of check standards are selected to monitor the range of items that are calibrated by the laboratory.

2   Summary

A check standard is obtained and calibrated several times initially to establish a reliable mean value and to estimate the standard deviation of calibration. All such calibrations are made using the applicable procedure that is used for calibration. A reference value may be obtained using a better calibration than the one that will be monitored. Directions for preparing and using a control chart for monitoring the mean ($\bar{x}$) and the standard deviation and range chart are given. The $\bar{x}$ control chart monitors the process with respect to both the standard and the variability, while the standard deviation or range chart monitors its short-term precision. When the calibration process is determined to be in a state of statistical control, the calibrations made at that time may be considered to be valid and the process standard deviation may be used, as appropriate, to calculate the uncertainty for the calibrations using SOP 29.

Note: If a full evaluation of the process bias is desired, it is best if the reference value of the check standard are provided by an outside laboratory accredited to perform the applicable calibrations.

3   Equipment

   3.1 A check standard is required, and should be constructed of similar materials and design as the standards under calibration.

   3.2 All equipment designated in the applicable SOP.

4   Procedure

   4.1 Initial Measurements

4.1.1 Calibrate the check standard a minimum of 12 times to establish the baseline chart. A calibration is defined as the result of duplicate measurements as required by the SOP (i.e., a complete test consists of two runs). Calibrations may be made on successive days, but no two complete tests should be made on any single day. Note: 25 to 30 points are recommended to determine uncertainties.

4.1.2 Tabulate the measurement data using the notation and a form such as the one contained in the Appendix of this SOP. The data may be maintained in a spreadsheet or other electronic program in lieu of a paper form.

4.1.3 A standard deviation may be calculated for each set of runs according to the appropriate SOP with a pooled standard deviation determined for the measurement process. This is preferred.

4.1.4 Calculate the mean of the two trials $\bar{x}_i$ and the ranges between runs. The ranges R_i, are the absolute differences between run 1 and run 2 for the n tests. Be sure that only absolute values are used in the determination of the range and average range!

4.1.5 Calculate the average range $|\bar{R}|$ of the trials, for the n tests as follows:

$$\bar{R} = \frac{\sum |R_i|}{n}$$ Eqn. 1

Estimate the standard deviation of the process s_p, for each set of made runs according to the SOP.

4.1.6 The standard deviation may be calculated using the average range as follows (obtain values for d_2^* from NISTIR 6969 Table 9.10 and see NISTIR 6969 Section 8.3 for additional notes):

$$s_p = \frac{\bar{R}}{d_2^*}$$ Eqn. 2

4.2 Construction of Control Charts (See also SOP 9)

4.2.1 Construct the following control charts using the data of section 4.1.

4.2.2 Construct an $\bar{x}$ control chart for a check standard with the following control limits:

Reference value (when available)
Central Line = $\bar{\bar{x}}$ (mean of the average values)
Lower warning limit (LWL) = $\bar{\bar{x}} - 2s_p$
Lower control (or action) limit (LCL) = $\bar{\bar{x}} - 3s_p$

Upper warning limit (UWL) = $\overline{\overline{x}} + 2s_p$

Upper control (or action) limit (UCL) = $\overline{\overline{x}} + 3s_p$

4.2.3 Construct a Standard Deviation or Range chart using the same approach. However, you may use 2 and 3 as the respective multipliers for the Upper Warning Limit and Upper Control Limits. Note that there will be no negative numbers when calculating standard deviations.

4.2.4 Construct an R (range chart) control chart for duplicate measurements having the following control (or action) limits. Note that R (the range) and $|d|$ (absolute difference of duplicate measurements) are equivalent for duplicate measurements.

Central Line = $\overline{R}$ (average range)

LCL = LWL = 0

(There should be no negative numbers recorded when using absolute values!)

UWL   =   2.512 $\overline{R}$

UCL   =   3.267 $\overline{R}$

4.2.5 These limits are t values for 95 % and 99.7 % confidence intervals for a sample size of 30.

4.2.6 The recommended format for construction of R control charts is given in NISTIR 6969, Section 7.4.

4.3 Use of Control Charts

4.3.1 An appropriate check standard is calibrated each time the laboratory performs calibrations using the SOP. If the calibrations extend over several days, the check standard is calibrated daily. The values of $\overline{x}$ and s_p or R for each calibration of the check standard are plotted on the respective control charts, preferably in sequential order. The limits on the charts are such that 95 % of the values should fall within the warning limits and rarely should a value fall outside of the control limits, provided that the system is in a state of statistical control.

4.3.2 If the plotted value of $\overline{x}$ lies outside of the control limits and the corresponding value on the standard deviation or range chart is within the control limits, a source of systematic error is suspected.

4.3.3 If the values for the standard deviation or range chart fall outside of the warning limits but inside of the control limits, a decrease in precision is indicated. Other problems should be investigated.

4.3.4 No calibration data should be accepted when the system is out of control.

4.3.5 If the plotted values for either $\bar{x}$, s_p or R are outside of the warning limits but inside of the control limits, a second set of duplicate calibrations should be made. If the new values are within the warning limits, the process may be considered to be in control. If they lie outside of the warning limits, lack of control is indicated. Corrective actions should be taken and attainment of control demonstrated before calibration measurements are considered to be acceptable.

4.3.6 Even while the system is in an apparent state of control, incipient troubles may be indicated when the control data show short- or long-term trends, shifts, or runs. The t-test and F-test may be used to assess the significance of such observations (see NIST IR 6969 Section 8.9, 8.10, and 8.11).

4.4 Interpretation of Control Chart Data

4.4.1 Demonstration of "in control" status indicates that the calibration process is consistent with the past experience of the laboratory. That is to say, there is no reason to believe that excessive systematic error or changes in precision have occurred.

4.4.2 To the extent appropriate, the precision of measurement of the check standard may be extended to the calibration of other standards of similar nominal size made by the same measurement method.

4.4.3 Extension of the s_p for the check standard to other calibrations assumes that all aspects of its calibration correspond to those for the other calibration.

Appendix
Control Chart Check Standard Data

Check Std ID _____   Nominal Value _____

Test Number	Date	Run 1	Run 2	Average of Runs	Range* $\|d\| = \|$Run 1 - Run 2$\|$ (Max – Min)	Standard deviation**
1						
2						
3						
4						
5						
6						
7						
8						
9						
10						
11						
12						
13						
14						
15						
SUM						
				$\sum \overline{x}$	$\sum \|d\|$	pooled std dev:

$n*** = $ _____

$\overline{R} = \dfrac{\sum \|R\|}{n} = $ _____   UWL $= 2.512\,\overline{R}\ =$

UCL $= 3.267\,\overline{R}\ =$

* This is the range, R, of the two trials and is actually the larger value minus the smaller value.
** Use of the standard deviation and pooled standard deviations are preferred to the use of range as an estimate of the standard deviation.
***n is the number of tests used to calculate the control limits.

SOP 18

Standard Operating Procedure for
Calibration of Graduated Neck-Type Metal Volumetric Field Standards
Volumetric Transfer Method[1]

1   Introduction

    1.1   Purpose of Test

        This procedure may be used to calibrate small non-pressurized, graduated neck-type, metal field standards such as the 5 gal (or 20 L) standards used by weights and measures officials to test liquid dispensing equipment, gasoline pumps, for example. The test measure prover should be evaluated for conformance to appropriate specifications if being used for legal weights and measures applications.

        The procedure assumes that the water temperature is stable during the transfer from the standard to the unknown test measure. SOP 19 is a more appropriate procedure when temperature corrections are needed due to lack of water equilibration, temperature differences between the standard and unknown provers, or unstable environments.

        Limiting factors: For a 5 gal test with a stainless steel standard and stainless steel unknown test measure, the temperature of the water between the standard and unknown must not change more than 0.5 °C during the calibration. If the unknown test measure is mild steel, the change in water temperature between standard and unknown must be less than 0.2 °C during the calibration. If these limits are exceeded, use SOP 19. This is to ensure that the impact on measured values is less than the resolution and repeatability on a 5 gal test measure with 1 in^3 graduations. If smaller graduations are present, error due to temperature variations must be evaluated further or SOP 19 is recommended.

    1.2   Prerequisites

        1.2.1   Verify that the unknown prover has been properly cleaned and vented with all petroleum products removed prior to submission for calibration to ensure laboratory safety.

        1.2.2   Verify that valid calibration certificates are available for all of the standards used in the test.

---

[1] Non-SI units are predominately in common use in State legal metrology laboratories, and/or the petroleum industry for many volumetric measurements, therefore non-SI units have been used to reflect the practical needs of the laboratories performing these measurements as appropriate. Most laboratory standards for this calibration procedure are 5 gal "slicker-plate" type standards. Very few laboratories have 20 L "slicker-plate" type standards.

1.2.3 Verify that the slicker-plate type standard to be used has sufficiently small standard uncertainties for the intended level of calibration (i.e., less than 0.2 in^3 with a 95 % confidence interval).

1.2.4 Verify the availability of an adequate supply of clean water (GLP 10).

1.2.5 Verify that the operator has had specific training and is proficient in SOP 18, GMP 3, SOP 17 and is familiar with the operating characteristics and conditioning of the standards used.

1.2.6 Verify that the laboratory facilities meet the following minimum conditions to enable meeting the expected uncertainty achievable with this procedure:

Table 1.   Laboratory environmental conditions.

Procedure	Temperature	Relative Humidity
Volume transfer	18 °C to 27 °C, stable to ± 2.0 °C/h	35 % to 65 % ± 20 % maximum change / 4 h

1.3 Field tests

1.3.1 A "field" calibration is considered one in which a calibration is conducted in an uncontrolled environment, such as out-of-doors. Calibrations conducted under field and laboratory conditions are not considered equivalent and uncertainties must reflect the conditions of the calibration.

1.3.2 The care required for field calibrations includes proper safety, a clean and air-free water supply, measurement control programs, and a stable temperature environment shaded from direct sunshine to allow the prover, field standard, and test liquid (water) to reach an equilibrium temperature with minimal evaporation. Environmental conditions must be selected to be within stated laboratory conditions during the measurements. All data and appropriate environmental conditions must be documented regardless of test location. SOP 19 is a more suitable procedure for non-laboratory conditions.

1.3.3 An increased number of check standard verifications are required to ensure continued suitability of calibration values generated in field conditions as well as to verify the validity of any standards taken out of a secure laboratory environment

2 Methodology

2.1   Scope, Precision, Accuracy

This procedure is applicable for the calibration of a small test measure within the limitations of the standards available. The precision attainable depends on the care used in the various volumetric adjustments and readings, in the strict observance of drainage times, and the internal cleanliness of the various volumetric vessels which can influence their drainage characteristics. The accuracy depends on the uncertainties of the calibrations of the standards used.

2.2   Summary

Water is delivered from the standard to the vessel under calibration. Because the "to deliver" volume of the latter is calibrated, the delivery must be into a "wet-down" vessel. The wet-down also ensures consistent retention in the slicker-plate type standard. The gauge scale of the test vessel is adjusted to a correct reading, as necessary, and then sealed.

2.3   Equipment

2.3.1   Calibrated slicker-plate standard made of stainless steel, with recent calibration certificate and demonstrated metrological traceability to the international system of units (SI), which may be to the SI through a National Metrology Institute such as NIST, and whose volume is equivalent to that of the vessel to be calibrated.

2.3.2   Calibrated thermometer, accurate to ± 0.1 °C, with recent calibration certificate and demonstrated metrological traceability to the international system of units (SI), which may be to the SI through a National Metrology Institute such as NIST.

2.3.3   Meniscus reading device. (See GMP 3).

2.3.4   Timing device (Calibration is not required; uncertainty of the measurement only needs to be less than 5 s for a 30 s pour time.)

2.3.5   Supply of clean water, preferably soft water (filtered if necessary).

2.4   Procedure

2.4.1   Cleanliness Verification - Fill and drain both standard and vessel to be calibrated and check for any soiling that would affect drainage, as evidenced by clinging droplets, greasy films, and the like. Clean either or both with non-foaming detergent and water, as necessary, and rinse thoroughly. (See GMP 6).

2.4.2 Fill a hand-held vessel with water to its nominal level and pour contents during a 30 ± 5 s period then drain for a 10 s period after cessation of flow. Touch off any adhering drop from the neck. If a stationary test measure is being calibrated, the valve is opened, the measure is emptied, followed by a 30 s drain time after the cessation of flow. This constitutes the "wet-down" condition. Filling the vessel from the slicker-plate type standard following the instructions in steps 2.4.3 and 2.4.4 will ensure that both the standard and vessel are properly "wet-down".

2.4.3 Run 1 - Fill the slicker-plate standard with water, raised by surface tension, slightly higher than the rim. Use the slicker plate to strike off excess water, checking to see that no air bubbles are entrained in the water during the leveling process.

2.4.4 Open the valve at the base at the same time as removing the slicker plate from the top of the standar to transfer water from the standard to the wet-down vessel. Allow a 30 s drain period after cessation of flow.

2.4.5 Level the vessel (or suspend it by its handle, if appropriate) and read the neck scale. Record the reading.

2.4.6 Adjust the graduated scale of the vessel as described in 3.3. Seal the scale adjustment device.

2.4.7 Run 2 - Make a duplicate determination, which should agree with the former within ± 0.02 % of the volume (± 0.2 in^3 for 5 gal test measure). The test measure or prover must be capable of repeating within 0.02 % of the test volume during calibration.

NOTE: If excess disagreement, check all vessels for cleanliness, leaks, or other damage, identifying and correcting any problems. Repeatability problems may be due to contamination or lack of cleanliness, or poor field conditions, such as when calibration is conducted in an unstable environment. Repeatability problems must be corrected before calibration can be completed.

3 Calculations

3.1 Because the water temperature is usually reasonably close to 60 °F, the coefficients of expansion of the standard and the test vessel are sufficiently close together, and the deliveries and readings are made over a short period of time, temperature corrections are not made in this procedure. When conditions are not reasonably close to 60 °F and temperature corrections are needed, use SOP 19. If prover volumes, errors and/or corrections are reported, use calculations provided in SOP 19.

3.2   Within the accuracy requirements, no corrections arising from dissimilarities of the standard and vessel are necessary. If differences are suspected, use SOP 19.

3.3   The reading of Run 1 is used to adjust the scale of the vessel, if necessary, to the correct reading, which is set at the calibrated volume of the slicker-plate standard at 60 °F. Record the adjusted value as the "as left" value. Run 2 will validate the setting. Alternatively, the average of Run 1 and Run 2 may be used with the adjustment made after Run 2. In that case, a validation run should be conducted to ensure correct setting of the gage plate.

Note: If the accuracy requirements necessitate a temperature correction, the temperature of the water must be measured in both the standard and the unknown and the calibration is made according to the procedure given in SOP 19.

3.4   Determine and report the volume of the test vessel as follows:

$$Prover\ volume = V_{Nom} + C_s - gauge\ reading \qquad \text{Eqn. 1}$$

where:

V_{Nom} =   Nominal Volume (taking care to match units)
C_s =   Correction from the calibration report for the slicker-plate standard

4   Measurement Assurance

4.1   Duplicate the process with a suitable check standard. See SOP 17, SOP 20 and SOP 30. Plot the check standard volume and verify it is within established limits OR   a
t-test may be incorporated to check the observed value against an accepted value. The mean of the check standard observations is used to evaluate bias and drift over time. Check standard observations are used to calculate the standard deviation of the measurement process which contributes to the Type A uncertainty components.

4.2   If a standard deviation chart is also used for measurement assurance, the standard deviation of each combination of Run 1 and Run 2 is calculated and the pooled (or average) standard deviation and may be used to estimate the short-term variability in the measurement process. The short-term standard deviation may be used to incorporate an F-test (Observed vs. Accepted) into the measurement process and represents the variability in condition of test measures submitted for calibration.

5   Assignment of Uncertainties

The limits of expanded uncertainty, U, include estimates of the standard uncertainty of the laboratory volumetric standards used, u_s, plus the standard deviation of the process, s_p, at the 95 % level of confidence. See SOP 29 for the complete standard operating procedure for calculating the uncertainty.

5.1   The standard uncertainty for the standard, u_s, is obtained from the calibration report. The combined standard uncertainty, u_c, is used and not the expanded uncertainty, U, therefore the reported uncertainty for the standard will usually need to be divided by the coverage factor k.

5.2   The standard deviation of the measurement process, s_p, is obtained from control chart for the check standard to reflect performance over time (See SOP 17 or 20, and SOP 30). The larger of the value from the standard deviation over time for a check standard from Section 4.1 or from the standard deviation chart taken from Section 4.2 should be used in the uncertainty calculations.

5.3   Other standard uncertainties usually included at this calibration level primarily include 1) uncertainties associated with the ability to read the meniscus, only part of which is included in the process variability due to parallax and visual capabilities, and 2) uncorrected temperature corrections related to the cubical coefficient of expansion for the prover under test, use of proper temperature corrections, and the accuracy of temperature measurements. Additional factors that might be included are: round robin data showing reproducibility, environmental variations over time, and bias or drift of the standard as noted in control charts.

5.4   To properly evaluate uncertainties and user requirements (tolerances), assessment of additional user uncertainties may be required by laboratory staff. Through proper use of documented laboratory and field procedures, additional uncertainty factors may be minimized to a level that does not contribute significantly to the previously described factors. Additional standard uncertainties in the calibration of field standards and their use in meter verification may include: how the prover level is established, how delivery and drain times are determined, the use of a proper "wet-down" prior to calibration or use, the cleanliness of the prover and calibration medium, prover retention characteristics related to inside surface, contamination or corrosion, total drain times, and possible air entrapment in the water.

6   Report

    6.1   Report results as described in SOP 1, Preparation of Calibration/Test Results, with the addition of the following:

        6.1.1   Volume, reference temperature, uncertainty, material, thermal coefficient of expansion (assumed or measured), construction, any identifying markings, tolerances (if appropriate), laboratory temperature, water temperature(s) at time of test, barometric pressure, relative humidity, and any out–of-tolerance conditions.

SOP 19

**Standard Operating Procedure for
Calibration of Graduated Neck-Type Metal Provers
(Volume Transfer Method)[1]**

1   Introduction

1.1   Purpose of Test

This procedure is used to calibrate graduated neck type metal test measures and provers (20 L (5 gal) and larger) that are used in verification of petroleum, biodiesel, ethanol, milk, and/or water meters. The test measure prover should be evaluated for conformance to appropriate specifications if being used for legal weights and measures applications.

1.2   Prerequisites

1.2.1   Verify that the unknown prover has been properly cleaned and vented, with all petroleum products removed prior to submission for calibration to ensure laboratory safety and compliance with environmental disposal requirements.

1.2.2   Verify that valid calibration certificates are available for all of the standards used in the test.

1.2.3   Verify that the standards to be used have sufficiently small standard uncertainties for the intended level of calibration.

1.2.4   Verify the availability of an adequate supply of clean water (GLP 10) (Note: this is critical when calibrating food-grade provers.)

1.2.5   Verify that the operator has had specific training and is proficient in SOP 18, SOP 19, SOP 17, SOP 20, GMP 3, and is familiar with the operating characteristics and conditioning of the standards used.

1.2.6   Verify that the laboratory facilities meet the following minimum conditions to make possible the expected uncertainty achievable with this procedure:

---

[1] Non-SI units are predominately in common use in State legal metrology laboratories, and/or the petroleum industry for many volumetric measurements, therefore non-SI units have been used to reflect the practical needs of the laboratories performing these measurements as appropriate. .

Table 1.   Laboratory environmental conditions.

Procedure	Temperature	Relative Humidity
Volume Transfer	18 °C to 27 °C, stable to ± 2.0 °C/h	35 % to 65 % ± 20 %, maximum change / 4 h

1.3   Field tests

  1.3.1   A "field" calibration is considered one in which a calibration is conducted in uncontrolled environments, such as out-of-doors. Calibrations conducted under field and laboratory conditions are *not* considered equivalent.

  1.3.2   The care required for field calibrations includes proper safety, a clean and air-free water supply, measurement control programs, and a stable temperature environment shaded from direct sunshine to allow the prover, field standard, and test liquid (water) to reach an equilibrium temperature with minimal evaporation. Environmental conditions must be selected to be within stated laboratory conditions during the measurements. . All data and appropriate environmental conditions must be documented regardless of test location.

2   Methodology

  2.1   Scope, Precision, Accuracy

  This procedure is applicable for the calibration of any size metal prover within the limitations of the standards available. The precision attainable will depend on strict adherence to the procedure, the care in volumetric adjustments, and the number of transfers, in the case of multiple transfers. . The accuracy will depend on the standards used.

  2.2   Summary

  Water is delivered from a volumetric standard to the prover being calibrated. Depending on the respective volumes, multiple transfers may be required. While these should be minimized, a maximum number of 15 transfers are permitted to ensure that final uncertainties and systematic errors are sufficiently small for the intended applications. The temperature cannot be considered to be constant during multiple transfers; hence, the temperature of the water for each transfer must be measured. Because of the large volumes, the difference in thermal expansion of the respective vessels must be considered.

  2.3   Equipment

2.3.1 Calibrated volumetric standard with recent calibration certificate and demonstrated metrological traceability to the international system of units (SI), which may be to the SI through a National Metrology Institute such as NIST.

2.3.2 Calibrated 1 gal flask (or other suitable size) to calibrate neck of prover and a funnel with demonstrated metrological traceability to the international system of units (SI), which may be to the SI through a National Metrology Institute such as NIST.

2.3.3 Meniscus reading device (See GMP 3).

2.3.4 Calibrated thermometer, accurate to ± 0.1 °C, with recent calibration certificate and demonstrated metrological traceability to the international system of units (SI), which may be to the SI through a National Metrology Institute such as NIST.

2.3.5 Timing device (calibration is not required; uncertainty of the measurement only needs to be less than 5 s for a 30 s pour time.)

2.3.6 Supply of clean water, preferably soft water (filtered if necessary).

2.3.7 Sturdy platform, with appropriate safety conditions, with sufficient height to hold standard and to permit transfer of water from it to the prover by gravity flow.

2.3.8 Clean pipe or tubing (hoses) to facilitate transfer of water from the laboratory standard to prover. Pipe and hose lengths must be minimized to reduce water retention errors. Care must be taken during wet-downs and runs to ensure complete drainage and consistent retention in all hoses or pipes.

2.4 Procedure

2.4.1 Cleanliness verification

Fill and drain both the standard and unknown test measure or prover to be calibrated and check for visual evidence of soiling and of improper drainage. If necessary, clean with non-foaming detergent and water (see GMP 6).

2.4.2 Neck scale plate verification

2.4.2.1 Fill the unknown prover with water from the standard. Check the prover level condition in the same way in which it will be used and adjust if necessary. Check the prover system for leaks. This is a wet-down run.

2.4.2.2 Bleed the liquid level down to a graduation near the bottom of the upper neck. "Rock" the prover to "bounce" the liquid level, momentarily, to ensure that it has reached an equilibrium level. . Read and record this setting to be used as the initial scale reading sr_i for verification of the neck scale plate.

2.4.2.3 Recheck the scale reading, then add water from calibrated standards equal to $^1/_4$ or $^1/_5$ of the graduated neck volume and record the scale reading.

2.4.2.4 Repeat 2.4.2.3 by successive additions until water is near the top of the scale. Record scale readings after each addition. . The last reading will be the final scale reading, sr_f. The closer the water is to the top of the neck, the harder it may be to "bounce" the liquid in the gauge.

2.4.2.5 A plot of scale readings with respect to the total volume of water that is added V_w should be linear and will be a gross check of the validity of this calibration.

2.4.2.6 Calculate and assess the accuracy of the neck scale for each interval. . The error should be less than 0.5 % of the graduated neck volume or $^1/_4$ of a graduation (whichever is smaller). . If more than this, the scale should ideally be replaced. Alternatively, a Neck Scale Correction Value (NSCV) may be issued with instructions to the user if it is anticipated that this correction value will be used.

2.4.2.7 The neck scale correction value is calculated as follows:

$$NSCV = \frac{V_w}{\left(sr_f - sr_i\right)}$$

Table 2.   Variables for neck scale correction value equation.

$NSCV$	Neck scale correction value
Vw	Total volume of water added to neck
sr_f	Scale reading, final
sr_i	Scale reading, initial

2.4.3   Body Calibration

2.4.3.1 Fill prover with water and level it. . Drain water, then wait 30 s after cessation of full flow, before closing drain valve. This establishes a "wet-down" condition for provers with no bottom zero. . If a bottom zero is present, follow the guidance provided in SOP 21 for LPG provers as follows: When the liquid reaches the top of the lower gage glass, close the valve and allow the water to drain from the interior of the prover into the lower neck for 30 s. . Then bleed slowly with the bleed valve (4) until the bottom of the liquid meniscus reaches the zero graduation. . (This step should be started during the 30 s drain period but should not be completed before the end of the drain period.)

Alternatively, the prover may be completely drained with a 30 s drain time and then refilled with a funnel and small volume of water to set the zero mark (which will add to the prover calibration uncertainty due to variable retention characteristics).

2.4.3.2 Run 1. Fill the standard and measure and record the temperature.

2.4.3.2.1 Measure and record the temperature of the water in the standard, t_1, then adjust the standard prover to its reference mark or record the neck reading, and then discharge into the unknown prover. Wait 30 s after cessation of full flow to attain specified drainage, then close the delivery valve.

2.4.3.2.2 Repeat step 2.4.3.2.1. as many times as necessary (note the 15-drop limit) to fill the unknown prover to its nominal level. Record the temperature of water in the standard for each drop, t_1 to t_N. Level the unknown prover as necessary and record the neck reading. Measure the temperature of the water in the unknown prover, t_x, and record. An average temperature, calculated from temperatures taken at multiple locations from within the unknown prover may be appropriate.

2.4.3.2.3 Perform the calculations described in section 3 to determine the prover volume at the appropriate reference temperature.

2.4.3.4 Adjust the scale as needed. If adjusted, record the adjusted prover gauge reading for determining the "as left" value for Run 1. Run 2 will validate the setting. Alternatively, the average of Run 1 and Run 2 may be used with the adjustment made after Run 2. In that case, a validation run should be conducted to ensure correct setting of the gage plate.

2.4.3.5 Run 2 - Repeat the process described in 2.4.3.2. The duplicate determination should agree with the former within ± 0.02 % of the volume. The test measure or prover must be capable of repeating within 0.02 % of the test volume during calibration.

NOTE: If excess disagreement, check all vessels for cleanliness, leaks, or other damage, identifying and correcting any problems. Repeatability problems may be due to contamination or lack of cleanliness, or poor field conditions, such as when calibration is conducted in an unstable environment. . Repeatability problems must be corrected before calibration can be completed.

2.4.3.7 Seal the equipment as specified in the laboratory policy.

3   Calculations

The following calculations assume that the standard was calibrated using a reference temperature of 60 °F (15.56 °C) and that you are calibrating a field standard to a reference temperature of 60 °F (15.56 °C). Equations for situations where different reference temperatures are involved will follow.

3.1   Single Delivery

3.1.1   Calculate V_{X60}, the volume of the unknown prover at 60 °F, using the following equation:

$$V_{X60} = \frac{\rho_1 \left\{ \left(V_{S60} + \Delta_1\right)\left[1 + \alpha\left(t_1 - 60\text{ °F}\right)\right]\right\}}{\rho_x \left[1 + \beta\left(t_x - 60\text{ °F}\right)\right]}$$ Eqn. 1

3.2   Multiple Deliveries

3.2.1   Calculate V_{X60}, the volume of the unknown prover at 60 °F, using the following equation:

$$V_{X60} = \frac{\rho_1\left\{\left(V_{S60} + \Delta_1\right)\left[1 + \alpha\left(t_1 - 60\text{°F}\right)\right]\right\} + \rho_2\left\{\left(V_{S60} + \Delta_2\right)\left[1 + \alpha\left(t_2 - 60\text{°F}\right)\right]\right\} + \dots + \rho_N\left\{\left(V_{S60} + \Delta_N\right)\left[1 + \alpha\left(t_N - 60\text{°F}\right)\right]\right\}}{\rho_x\left[1 + \beta\left(t_x - 60\text{°F}\right)\right]}$$ Eqn. 2

Table 3.   Variables for V_{X60} equations.

Symbols Used in Equations	
V_{X60}	volume of the unknown vessel at 60 °F
V_{S60}	volume of the standard vessel at 60 °F
$\rho_1, \rho_2, \dots, \rho_N$	density of the water in the standard prover where ρ_1 is the density of the water for the first delivery, ρ_2 is the density of the water for the second delivery, and so on until all N deliveries are completed

$\Delta_1, \Delta_2,...., \Delta_N$	volume difference between water level and the reference mark on the standard where the subscripts 1, 2,....,N, represent each delivery as above. If the water level is below the reference line, Δ is negative. . If the water level is above the reference line, Δ is positive. If the water level is at the reference line, Δ is zero NOTE: units must match volume units for the standard.
$t_1, t_2, ..., t_N$	temperature of water for each delivery with the subscripts as above
α	coefficient of cubical expansion for the standard in units / °F
β	coefficient of cubical expansion for the prover in units / °F
t_x	temperature of the water in the filled unknown vessel in units °F
ρ_x	density of the water in the unknown vessel in g/cm^3
Note: Values for the density of water at the respective temperatures may be found in Table 9.8 (in NISTIR 6969) or it may be calculated from the equation given in GLP 10.	

3.3 Prover Error/Correction or Deviation From Nominal

The total calculated volume of the prover at its reference temperature should be reported on the calibration report.

The prover volume for an open neck prover equals the V_{x60} value minus the gauge reading that is the difference from the nominal volume (with matched units).

$$Prover\ volume = V_{X60} - gauge\ reading \qquad \text{Eqn. 3}$$

$$Prover\ error = Prover\ volume\ - V_{Nom} \qquad \text{Eqn. 4}$$

$$Prover\ error = V_{X60} - gauge\ reading\ - V_{Nom} \qquad \text{Eqn. 5}$$

where:
   V_{Nom} = Nominal Volume (taking care to match units)

V_{X60} is the calculated volume of water that should be observed in the prover. A positive prover error means that the prover is larger than nominal. A negative prover error means that the prover is smaller than nominal.

Example 1:   If V_{X60} is 100.02 gal and gauge reading is 0.02 gal (above nominal);
          then the prover volume at nominal is 100.00 gal;
          and the prover error and correction are 0; and
          no adjustment is needed.

Example 2:   If V_{X60} is 100.02 gal and gauge reading is -0.02 gal (below nomival);
          then the prover volume at nominal is 100.04 gal;
          the prover error is + 0.04 gal; and

to adjust the prover, set the gauge to read 0.02 gal (the volume level will show a gauge reading of 0.02 gal, which is 4.62 in^3 or about 5 in^3, above nominal.)

3.4   Alternative Reference Temperatures

3.4.1 Reference temperatures other than 60 °F (15.56 °C) may occasionally be used. Common reference temperatures for other liquids follow:

Commodity	Reference Temperature
Frozen food labeled by volume (e.g., fruit juice)	-18 °C (0 °F)
Beer	3.9 °C (39.1 °F)
Food that must be kept refrigerated (e.g., milk)	4.4 °C (40 °F)
Distilled spirits or petroleum	15.56 °C (60 °F)
Petroleum (International Reference)	15 °C (59 °F)
Wine	20 °C (68 °F)
Unrefrigerated liquids (e.g., sold unchilled, like soft drinks)	20 °C (68 °F)
Petroleum (Hawaii)	26.67 °C (80 °F)

Equations for calculations when using alternative reference temperatures follow:

3.5   Single Delivery

3.5.1   Calculate V_{Xtref}, the volume of the unknown prover at its designated reference temperature (°F), using the following equation:

$$V_{Xtref} = \frac{\rho_1\left\{\left(V_{S\,tref} + \Delta_1\right)\left[1+\ \alpha\left(t_1 - t_{ref\,S}\right)\right]\right\}}{\rho_x\left[1+\beta\left(t_x - t_{ref\,X}\right)\right]} \qquad \text{Eqn. 6}$$

3.6   Multiple Deliveries

3.5.1   Calculate V_{Xtref}, the volume of the unknown prover at its designated reference temperature, using the following equation:

$$V_{Xtref_X} = \frac{\rho_1\left\{\left(V_{Stref_S}+\Delta_1\right)\left[1+\alpha_1\left(t_1 - t_{refS}\right)\right]\right\}+\rho_2\left\{\left(V_{Stref_S}+\Delta_2\right)\left[1+\alpha_2\left(t_2-t_{refS}\right)\right]\right\}+\ldots+\rho_N\left\{\left(V_{Stref_S}+\Delta_N\right)\left[1+\alpha_N\left(t_N-t_{refS}\right)\right]\right\}}{\rho_x\left[1+\beta\left(t_x - t_{ref_X}\right)\right]} \qquad \text{Eqn. 7}$$

Table 3A.   Variables for V_{Xtref_X} equations.

Symbols Used in Equations	
V_{Xtref_X}	volume of the unknown vessel, V_X at its designated reference temperature, t_{ref_X}
V_{Stref_S}	volume of the standard vessel, V_S at its designated reference

	temperature, $t_{ref}s$
$\rho_1, \rho_2,..., \rho_N$	density of the water in the standard where ρ_1 is the density of the water for the first delivery, ρ_2 is the density of the water for the second delivery, and so on until all N deliveries are completed
$\Delta_1, \Delta_2,..., \Delta_N$	volume difference between water level and the reference mark on the standard where the subscripts 1, 2,..., N, represent each delivery as above. If the water level is below the reference line, Δ is negative. If the water level is above the reference line, Δ is positive. If the water level is at the reference line, Δ is zero. NOTE: units must match volume units for the standard.
$t_1, t_2, ..., t_N$	temperature of water for each delivery with the subscripts as above
α	coefficient of cubical expansion for the standard in its designated units
β	coefficient of cubical expansion for the prover in its designated units
t_x	temperature of the water in the filled unknown vessel in designated units
ρ_x	density of the water in the prover in g/cm^3

Note: Values for the density of water at the respective temperatures may be found in Table 9.8 (in NISTIR 6969) or it may be calculated from the equation given in GLP 10. Note: The cubical coefficient of the materials must match the unit assigned to the temperature measurement.

4 Measurement Assurance

4.1. Duplicate the process with a suitable check standard or have a suitable range of check standards for the laboratory. .See SOP 17, SOP 20 and SOP 30. .Plot the check standard volume and verify it is within established limits OR a t-test may be incorporated to check the observed value against an accepted value. The mean of the check standard observations is used to evaluate bias and drift over time. Check standard observations are used to calculate the standard deviation of the measurement process which contributes to the Type A uncertainty components.

4.2. If a standard deviation chart is used for measurement assurance, the standard deviation of each combination of Run 1 and Run 2 is calculated and the pooled (or average) standard deviation is used as the estimate of variability in the measurement process. Note: the pooled or average standard deviation over time will reflect varying conditions of test items that are submitted to the laboratory.

5 Assignment of Uncertainties

5.1 The limits of expanded uncertainty, U, include estimates of the standard uncertainty of the laboratory volumetric standards used, u_s, plus the standard deviation of the process,

s_p, at the 95 % level of confidence. See SOP 29 for the complete standard operating procedure for calculating the uncertainty.

5.1.1   The standard uncertainty for the standard, u_s, is obtained from the calibration report. The combined standard uncertainty, u_c, is used and not the expanded uncertainty, U, therefore the reported uncertainty for the standard will usually need to be divided by the coverage factor k. See SOP 29 for the complete standard operating procedure for calculating the uncertainty when multiple deliveries or multiple standards are used. Fifteen is the maximum recommended number of deliveries from a laboratory standard to a prover under test to minimize calibration uncertainties to the levels identified previously.

5.2   Neck calibration uncertainty should be estimated based on the uncertainty of standards used, errors observed during calibration and the repeatability of the neck calibration.

5.3   The standard deviation of the measurement process s_p, is taken from control chart for a check standard or from standard deviation charts from provers of similar size (See SOP 17, SOP 20 and SOP 30).

5.4   Other standard uncertainties usually included at this calibration level primarily include 1) uncertainties associated with the ability to read the meniscus, only part of which is included in the process variability due to parallax and visual capabilities, and 2) uncertainties associated with temperature corrections that include values for the cubical coefficient of expansion for the prover under test, the accuracy of temperature measurements, and factors associated with potential gradients in measuring the temperature in test measures or provers. Additional factors that might be included are: round robin data showing reproducibility, environmental variations over time, and bias or drift of the standard as noted in control charts.

5.4.1   To properly evaluate uncertainties and user requirements (tolerances), assessment of additional user uncertainties may be required by laboratory staff. Through proper use of documented laboratory and field procedures, additional uncertainty factors may be minimized to a level that does not contribute significantly to the previously described factors. Additional standard uncertainties in the calibration of field standards and their use in meter verification may include: how the prover level is established, how delivery and drain times are determined, the use of a proper "wet-down" prior to calibration or use, whether gravity drain is used during calibration or whether the volume of water is eliminated by pumping, the cleanliness of the prover and calibration medium, prover retention characteristics related to inside surface, contamination or corrosion, and total drain times, and possible air entrapment in the water, and connecting pipes. Systematic errors may be observed between laboratory calibration practices where a gravity drain is used and field use where the pumping system is used.

6   Report

   6.1   Report results as described in SOP 1, Preparation of Calibration/Test Results, with the addition of the following:

   Prover volume, reference temperature, uncertainty, material, thermal coefficient of expansion (assumed or measured), construction, any identifying markings, tolerances (if appropriate), laboratory temperature, water temperature(s) at time of test, barometric pressure, relative humidity, and any out–of-tolerance conditions.

Additional References:

Bean, V. E., Espina, P. I., Wright, J. D., Houser, J. F., Sheckels, S. D., and Johnson, A. N., NIST Calibration Services for Liquid Volume, NIST Special Publication 250-72, National Institute of Standards and Technology, Gaithersburg, MD, (2006).

SOP 20

Standard Operating Procedure
for
Standard Deviation and Range Charts

1   Introduction

1.1   Purpose

This procedure describes a process to be followed to monitor the statistical control of a measurement process using standard deviation charts or range charts for any calibration method where it is not practical or feasible to maintain laboratory check standards.

1.2   Prerequisites

The procedure must match the calibration procedure that is being monitored.

2   Summary

Because of the size and cost of some laboratory standards, it is not always practical to have check standards remain in the laboratory for the purposes of measurement control. However, it is practical to maintain standard deviation (s_p) charts or range (R) charts for each nominal size standard to evaluate the standard deviation of replicate runs. Directions for preparing and using an R control chart that monitors the precision of the test procedure are given. It is assumed that standards of the same nominal capacity and design will have similar characteristics with respect to the repeatability of tests. Since it is not practical to run a sufficient number of tests on each unknown standard to determine the repeatability, the absolute difference between replicate test results on are graphed on the same s_p or R chart to reflect the repeatability of measurement of the unknown standards tested in the laboratory. Note: Provers of similar readability may be grouped together on the same chart (e.g., provers greater than and equal to 500 gal generally have 20 in^3 graduations and similar repeatability.)

3   Equipment

All equipment is designated in the respective calibration SOP.

4   Procedure

4.1   Data Collection

4.1.1   Conduct a minimum of two runs on the unknown test item per the designated SOP. A minimum of 12 sets of data must be available before a reasonably adequate data base is established. Note: 25 to 30 replicate sets of data are recommended to determine valid uncertainties.

4.1.2   Tabulate the measured errors as determined by each of the two trials using a form such as the one contained in the Appendix. The data may be maintained in a spreadsheet or other electronic program in lieu of a paper form. (If the unknown item is adjusted after the first trial to indicate zero error the first trial reading is evaluated after the adjustment.)

4.1.3   A standard deviation may be calculated for each set of runs according to the appropriate SOP with a pooled standard deviation determined for the measurement process. This is preferred.

4.1.4   Calculate the absolute difference $|d|$ of the two trials and the summation $\Sigma|d|$. Note that $|d| = R$, the range of the two trials. Be sure that only absolute values are used in the determination of the range and average range!

4.1.5   Calculate the average range of the trials, $\overline{R}$, for the n tests as follows:

$$\overline{R} = \frac{\Sigma|d|}{n}$$ Eqn. 1

4.1.6   The estimated standard deviation may be calculated using the average range as follows (obtain values for d_2^* from NISTIR 6969 Table 9.10 and see NISTIR 6969 Section 8.3 for additional notes):

$$s_p = \frac{\overline{R}}{d_2^*}$$ Eqn. 2

4.2   Construct Appropriate Charts

4.2.1.   Construct Standard Deviation Charts and Range Charts using the same approach. However, you may use 2 and 3 as the respective multipliers for the Upper Warning Limit and Upper Control Limits. Note that there will be no negative numbers when calculating standard deviations.

4.2.2.   Construct an R control chart having the following limits:

Central Line $= \overline{R}$

Lower control and warning limits LCL = LWL = 0
(There should be no negative numbers recorded when using absolute values!)
Upper warning limit UWL $= 2.512\,\overline{R}$
Upper Control limit UCL $= 3.267\,\overline{R}$

These limits are t values for 95 % and 99.7 % confidence intervals for a sample size of 30.

    4.2.3 The recommended format for construction of R control charts is given in NISTIR 6969, Section 7.4.

4.3   Use of Control Charts

    4.3.1 Two trials are run on each prover submitted to the laboratory for calibration. The values for standard deviation or range are plotted on the appropriate control chart, preferably in sequential order. The limits of the charts are such that 95 % of the values should fall within the warning limits and rarely should a value fall outside the control limits, provided that the system is in a state of statistical control.

    4.3.2 If the values plotted on the standard deviation or range chart fall outside of the control limit, a decrease in precision is indicated. Problems with the standards or process will need to be investigated.

    4.3.3 No calibration data should be accepted when the system is out of control.

    4.3.4 If a plotted value for standard deviation or range is outside of the warning limit but inside the control limit, a second set of duplicate calibrations should be made. If the new value for R is within the warning limit, the process may be considered in control. If it lies outside of the warning limit, lack of control is indicated. Corrective action should be taken and attainment of control demonstrated before calibration measurements are considered to be acceptable.

    4.3.5 Even while the system is in an apparent state of control, incipient troubles may be indicated when the control data show short- or long-term trends, shifts, or runs.

5   Interpretation of Control Chart Data

    5.1 Demonstration of "in control" indicates that the calibration process is consistent with the past experience of the laboratory. That is to say, there is no reason to believe that excessive changes in precision have occurred.

    5.2 The accuracy is inferred from a consideration of control of the sources of bias.

    5.3 To the extent appropriate, the precision of measurement of standards may be extended to the calibration of other standards of similar type, capacity and design.

Appendix
Standard Deviation and Range Data

Standard Capacity: : _____   Laboratory: _____

Test Number	Date	Run 1	Run 2	Average of Runs	Range* $\|d\| = \|$Run 1 - Run 2$\|$ (Max – Min)	Standard deviation**
1						
2						
3						
4						
5						
6						
7						
8						
9						
10						
11						
12						
13						
14						
15						
SUM						
				$\sum \bar{x}$	$\sum \|d\|$	

$n^{***} = $ _____

$$\bar{R} = \frac{\sum \|d\|}{n} = \underline{\hspace{3cm}}$$

UWL $= 2.512 \ \bar{R} = $ _____   UCL $= 3.267 \ \bar{R} = $ _____

* This is the range, R, of the two trials and is actually the larger value minus the smaller value.
** Use of the standard deviation and pooled standard deviations are preferred to the use of Range as an estimate of the standard deviation.
***n is the number of tests used to calculate the control limits.

SOP 21

**Standard Operating Procedure for
Calibration of LPG Provers[1]**

1   Introduction

   1.1   Purpose of Test

      This procedure may be used to calibrate a volume standard used to test systems designed to measure and deliver liquefied petroleum gas (LPG) in the liquid state by definite volume, whether installed in a permanent location or mounted on a vehicle. A schematic diagram of such a prover is shown in Figure 1, together with numbers, e.g., 1, 2, 3, and 4 to clarify the various operations described in the procedure. The parts labeled A, B, and C are hose connections used in meter testing (versus prover calibrations).

   1.2   Prerequisites

     1.2.1 Verify the unknown prover has been properly cleaned and vented with all petroleum products removed prior to submission for calibration to ensure laboratory safety.

     1.2.2 Verify that valid calibration certificates are available for the standard used in the test.

     1.2.3 Verify that the standards to be used have sufficiently small standard uncertainties for the intended level of the calibration.

     1.2.4 Verify the availability of an adequate supply of clean water (GLP 10).

     1.2.5 Verify that the operator has had specific training in SOP 17, SOP 18, SOP 19, SOP 20, SOP 21, and GMP 3.

     1.2.6 Cylinder of nitrogen or compressed air and a proper pressure regulator.

     1.2.7 Verify that the laboratory facilities meet the following minimum conditions to make possible the expected uncertainty achievable with this procedure:

---

[1] Non-SI units are predominately in common use in State legal metrology laboratories, and/or the petroleum industry for many volumetric measurements, therefore non-SI units have been used to reflect the practical needs of the laboratories performing these measurements as appropriate. The majority of LPG provers in use are 20 gal, 25 gal, and 100 gal nominal sizes. The volume of LPG provers is established at 60 °F and 100 psig.

Table 1.   Laboratory environmental conditions.

Procedure	Temperature	Relative Humidity
Volume Transfer	18 °C to 27 °C, stable to ± 2.0 °C/h	35 % to 65 % ± 20 %, maximum change / 4 h

1.3   Field tests

1.3.1   A "field" calibration is considered one in which a calibration is conducted in an uncontrolled environment, such as out-of-doors. Calibrations conducted under field and laboratory conditions are not considered equivalent.

1.3.2   The care required for field calibrations includes proper safety, a clean and air-free water supply, measurement control programs, and a stable temperature environment shaded from direct sunshine to allow the prover, field standard, and test liquid (water) to reach an equilibrium temperature with minimal evaporation. Environmental conditions must be selected to be within stated laboratory conditions during the measurements. All data and appropriate environmental conditions must be documented regardless of test location.

2   Methodology

2.1   Scope, Precision, Accuracy

This procedure is applicable for the calibration of LPG provers with capacities of 100 L to 500 L (20 gal to 100 gal) or larger when appropriate. Provers of the latter capacity (gal and in^3 units) are encountered most frequently, hence the procedure is written with that in mind. The changes necessary for testing provers of other capacities will be obvious and are not described in this document. The agreement of duplicate measurements made within a short period of time on a given 100 gal prover should be within 5 in^3 (0.02 gal). The accuracy will depend on the uncertainty in the volume of the standard, on the care exercised in making the various measurements and temperature readings, and on correct application of the corresponding corrections.

2.2   Summary

The procedure is a modification of one described by M.W. Jensen in NBS Handbook 99, "Examination of Liquefied Petroleum Gas Liquid- Measuring Devices." The LPG prover is calibrated with a known volume of water delivered into it from a standard prover of calibrated volume. Depending on the respective volumes, multiple transfers may be required. While these should be minimized, a maximum number of 15 transfers are permitted to ensure that final calibration uncertainties are sufficiently small to meet user applications. The LPG prover is pressurized and the liquid level is measured at each of several values of applied pressure. The calibration thus defines the capacity of the prover over its expected range of operational pressure.

2.3   Equipment

2.3.1   Calibrated standard prover of minimum volume of 10 gal but preferably of the same volume as the LPG prover, with recent calibration certificate and demonstrated metrological traceability to the international system of units (SI), which may be to the SI through a National Metrology Institute such as NIST.

2.3.2   A funnel and a calibrated 1 gal flask (or other suitable size) with recent calibration certificate and demonstrated metrological traceability to the international system of units (SI), which may be to the SI through a National Metrology Institute such as NIST, to calibrate the neck of prover.

2.3.3   Thermometers (2) accurately calibrated to 0.1 °C, with recent calibration certificate and demonstrated metrological traceability to the international system of units (SI), which may be to the SI through a National Metrology Institute such as NIST.

2.3.4   Meniscus reading device (See GMP 3).

2.3.5   Timing device (calibration is not required; uncertainty of the measurement only needs to be less than 5 s for a 30 s pour time.)

2.3.6   Supply of clean water, preferably soft water (filtered if necessary) (GLP 10).

2.3.7   Sturdy platform, with appropriate safety conditions, with sufficient height to hold standard and to permit transfer of water from it to the prover by gravity flow.

2.3.8   Clean the pipe or tubing (hoses) to facilitate transfer of water from the laboratory standard to prover. Nearly all LPG provers require reducers to be used between normal laboratory piping and the top hole on the prover. Pipe and hose lengths must be minimized to reduce water retention errors. Care must be taken during wet-downs and runs to ensure complete drainage and consistent retention in all hoses or pipes.

2.3.9   Compressed nitrogen cylinder or air, suitable regulator, and an appropriate pressure gauge. The calibration relies on the accuracy of the pressure gauge on the prover. It assumes that systematic errors in the prover pressure gauge will be present in field applications as well, thus calibration of the laboratory pressure gauge is *not* essential.

2.4   Procedure

2.4.1   Preliminary Operations

2.4.1.1   Install and level the standard(s) on a raised platform with appropriate security and safety ensured for the prover(s) and operator(s). Provide pipe or tubing for delivery of water by most direct route to prover.

2.4.1.2   Position and level the unknown prover where it can be reached from the elevated standard by the shortest feasible delivery system.

2.4.1.3   Remove the plug and relief valve (1) from the top, and extend the pipe into the hole. This may require the use of a reducer and a short length of hose (about 1 inch in diameter). If this is a tight fit, open the vapor return line valve (2) to provide an air bleed.

Use the prover inlet line (3) as a gravity drain. If necessary, remove the fitting on the end and connect a hose or pipe to make the necessary drain line.

Warning: ensure that a check valve is not plumbed into the prover inlet line. If it is, remove the check valve, otherwise the prover will need to be drained via the plug in the bottom of the lower neck.

2.4.2   Cleanliness Check

Both the standard and the unknown prover must be internally clean. This should be verified by checking that water drains properly from them. If necessary, either or both should be cleaned with water and non-foaming detergent (see GMP No. 6) to attain good drainage characteristics. Additional effort may be required to eliminate scaling from the inside of LPG provers.

2.4.3   Neck scale plate verification

2.4.3.1   Fill the unknown prover with water from the standard. Check the prover levels and adjust if necessary. Check the prover system for leaks. This is a wet-down run.

2.4.3.2   Bleed the liquid level down to a graduation near the bottom of the upper neck. "Rock" the prover to "bounce" the liquid level, momentarily, to ensure that it has reached an equilibrium level. Read and record this setting to be used as the initial scale reading sr_i. This is in preparation for calibration of the neck scale.

2.4.3.3   Remove the fill hose or pipe from the top and insert a funnel.

2.4.3.4   Recheck the scale reading, then add 1 gal of water (or a volume equal to approximately ¼ or $^1/_5$ of the graduated neck volume) from a suitable standard; record the scale reading.

2.4.3.5   Repeat 2.4.3.4 by successive additions until water is near the top of the scale (the neck capacity is usually about 5 gal). Record scale readings after each addition. The last reading will be the final scale reading, sr_f. The closer the water is to the top of the neck, the harder it may be to "bounce" the liquid in the gauge.

            A plot of scale readings with respect to the total volume of water that is added V_w should be linear and will be a gross check of the validity of this calibration.

2.4.3.6   Calculate and assess the accuracy of the neck scale for each interval. The error should be less than 0.5 % of the graduated neck volume. If more than this, the scale should ideally be replaced. Alternatively, a Neck Scale Correction Value (NSCV) may be issued with instructions to the user if it is anticipated that this correction value will be used.

2.4.3.7   The neck scale correction value is calculated as follows:

$$NSCV = \frac{V_w}{\left(sr_f - sr_i\right)} \qquad \text{Eqn. 1}$$

Table 2.   Variables for neck scale correction value equation.

NSCV	Neck scale correction value
V_w	Total volume of water added to neck
sr_f	Scale reading, final
sr_i	Scale reading, initial

2.4.4   Body Calibration

2.4.4.1   Drain the prover through its inlet valve and the liquid bleeder valve. When the liquid reaches the top of the lower gauge glass, close the inlet valve and allow the water to drain from the interior of the prover into the lower neck for 30 s, while controlling the water flow and level with the bleeder valve until the liquid meniscus reaches the zero graduation. The liquid level should be exactly at the zero graduation and the bleeder valve closed simultaneously at the 30 s drain time. (Draining with the bleeder valve close to the zero mark should be started during the 30 s drain period but should not be completed before the end of the drain period.)

Alternatively, though not recommended, the prover may be completely drained with a 30 s drain time and then refilled with a funnel that has been wet down, and small volume of water to set the zero mark. Errors from this process will result due to the additional drain and retention time and other factors associated with the process.

2.4.4.2   Transfer the volume from the standard in the usual manner, and record the standard and prover temperature readings. If multiple transfers are required, record temperature of the standard at the time of each transfer, but that of the prover only after the final transfer. "Rock" the prover to "bounce" the liquid in the upper gauge glass before reading. Record the scale reading after the nominal volume has been transferred into the unknown LPG prover.

Note regarding temperature measurements: A digital temperature sensing device with a long cable can allow insertion of the probe into the standard and the unknown to enable direct liquid measurements at the bottom, middle, and top of the provers. If the prover thermometer wells are used, ensure that the prover has equalized with the temperature of the water.

2.4.4.3   Drain the LPG prover as described in 2.4.4.1 and make another test run. Record the temperatures of the standard(s) and the unknown and final scale reading. Calculate the prover error at 0 psig for each run using the appropriate equations in Section 3; these values are used to evaluate repeatability of the test only. The error at 0 psig for the test runs should agree within 0.02 % of the prover volume or approximately one-half the prover tolerance (i.e., 5 in^3 on a 100 gal LPG prover). If the repeatability is larger than 0.02 % of the prover volume, continue until replicates agree within these limits, taking care to ensure that poor cleanliness, contamination, bubbles in hoses, leaking valves or seals are not contributing to poor repeatability. Repeatability problems may also be due to poor field conditions, such as when calibration is conducted in unstable environments. Repeatability problems must be corrected before calibration can be completed.

2.4.4.4   Replace the relief valve and plug in the top of the prover using suitable pipe joint compound or tape. Allow water from 2.4.4.3 (the second run) to remain in prover.

2.4.5   Prover Adjustments

2.4.5.1   The internal pressure and hence the volume of the prover may vary during use. Accordingly, a pressure correction must be made using the data of steps 2.4.6.

To minimize the amount of correction needed when the prover is in use, the prover should be adjusted to indicate its nominal capacity when 100 psig is applied. (An internal pressure of 100 psig is suggested as being convenient.) If the actual volume of the prover is not near a convenient whole gallon value and cannot easily be adjusted to a whole gallon value, a prover correction value can be computed (see 3.4) and added to the pressure correction values to obtain a set of combined prover and pressure correction values to be computed. The pressure correction is computed in 3.4.

2.4.5.2   Use a cylinder of nitrogen or compressed air and a proper pressure regulator with an integrated pressure gauge.

Connect the cylinder regulator output to the vapor return fitting (2) near the top of the neck. This may require fashioning a connection with steel pipe nipples or other appropriate materials and the existing fittings.

Caution: Ensure that all piping and fittings are rated for the pressures to which they will be exposed.

Make sure all valves are closed except the vapor return valve. Verify that the final scale reading has not changed since it was recorded (if it has changed it may signal a leak in one of the valves or fittings), and then slowly introduce pressure until the installed prover gauge reads 100 psig. Lightly tap the gauge to ensure that the gauge needle is not sticking. "Bounce" the liquid in the neck, then read and record the liquid level at this applied pressure.

Return pressure to 0 psig and reapply pressure as above. This reading at 100 psig should agree with the above recorded reading within 0.02 % of the nominal volume of the prover. Leaking seals or valves may cause problems with repeatability of the gauge readings under pressure.

2.4.5.3   With the pressure in the prover at 100 psig, adjust the upper scale to read the nominal volume. This is accomplished by adjusting the upper scale so that the water level reading is:

$$\text{Desired scale reading} = V_{X60} - (V_{NOM} * 1.00032) \qquad \text{Eqn. 2}$$

Take care to use like units in the calculation. The calculation for V_{X60} is given in 3.1, Eqn 5.

Note: The correction factor 1.00032 corrects for the compressibility of the water at 60 °F and 100 psig. If the upper scale is not adjustable, see 2.4.5.4.

2.4.5.4   For provers with only an adjustable lower scale (or one in which the upper scale is not adjustable), a prover correction, L_C, may be calculated at 100 psig as follows:

$$L_C = V_{X60} - (V_{NOM} * 1.00032) - s\,r_u \qquad \text{Eqn. 3}$$

where:

L_C	Prover correction at 100 psig
V_{X60}	Volume of unknown prover at 60° F
V_{NOM}	Nominal volume of prover
sr_u	Upper scale reading at 100 psig
1.00032	Correction factor for the compressibility of water at 100 psig

Take care to use like units in the calculation.

If the prover correction is negative, move the bottom scale down to increase the prover volume. If the prover correction is positive, move the bottom scale up to decrease the prover volume. The distance h that the bottom scale is to be moved is:

$$h = \frac{4\,|L_C|}{\pi\,d^2} \qquad\qquad \text{Eqn. 4}$$

where:

h	Distance in inches the bottom scale is to be moved, up or down
L_C	Prover correction at 100 psig in cubic inches
d	Inside diameter of the lower neck of the prover in inches (as noted on identification plate)

2.4.6   Pressure Correction

2.4.6.1   Return pressure to 0 psig. Record the reading. Slowly introduce pressure until the installed prover gauge reads 50 psig. Lightly tap the gauge to ensure that the gauge needle is not sticking. "Bounce" the liquid in the neck, then read and record the liquid level at this applied pressure.

2.4.6.2   Repeat step 2.4.6.1 at 100, 150, and 200 psig. Other pressure points in between those listed may be tested if so desired. (The water level should decrease 10 in^3 to 15 in^3 for each 50 psig increase in pressure, although this varies depending on the geometry and age of the LPG prover. Water level changes significantly greater than indicated may be due to leaking seals and/or valves.) Erratic pressure readings may also be due to air entrapment based on prover design; repeated pressurizing of the prover should eliminate entrapped air. Air entrapment problems due to design may need to be investigated and corrected.

2.4.6.3   Repeat step 2.4.6.1 as the pressure is bled down to 150, 100, 50, and 0 psig (atmospheric pressure). The readings should agree with those previously obtained within approximately 0.02 % of the nominal prover volume. If the data are not linear with respect to pressure, repeat the series of measurements above to verify the nonlinearity of the readings. Leaking seals or valves may cause problems with repeatability of the gauge readings under pressure. The cause of repeatability errors must be identified and corrected before continuing the calibration.

2.4.7   Final Operations

2.4.7.1   Seal the bottom and top scales as specified by laboratory policy and as appropriate.

2.4.7.2   Drain prover, then remove plug (5) at the lower neck to facilitate drainage below the lower gauge. If time permits, let the prover drain overnight.

2.4.7.3   With the nitrogen cylinder or compressed air connected, blow nitrogen or air through the prover to remove remaining moisture. Be sure to blow out the drain line and any other portions of the system that may have become contaminated with water.

2.4.7.4   If water has entered the pump-off system, pour several gallons of alcohol into the prover and pump the alcohol through the system to remove the water to prevent it from freezing in the pump when LP gas is used.

3   Calculations

3.1   Single Delivery

3.1.1   Calculate V_{X60}, the volume of the unknown prover at 60 °F, using the following equation:

$$V_{X60} = \frac{\rho_1 \left\{ \left(V_{S60} + \Delta_1 \right) \left[1 + \alpha \left(t_1 - 60\,°F \right) \right] \right\}}{\rho_x \left[1 + \beta \left(t_x - 60\,°F \right) \right]} \quad \text{Eqn. 5}$$

3.2   Multiple Deliveries

3.2.1   Calculate V_{X60}, the volume of the unknown prover at 60 °F, using the following equation:

$$V_{X60} = \frac{\rho_1 \left\{ \left(V_{S60} + \Delta_1 \right) \left[1 + \alpha \left(t_1 - 60°F \right) \right] \right\} + \rho_2 \left\{ \left(V_{S60} + \Delta_2 \right) \left[1 + \alpha \left(t_2 - 60°F \right) \right] \right\} + \ldots + \rho_N \left\{ \left(V_{S60} + \Delta_N \right) \left[1 + \alpha \left(t_N - 60°F \right) \right] \right\}}{\rho_x \left[1 + \beta \left(t_x - 60°F \right) \right]} \quad \text{Eqn. 6}$$

Table 3.   Variables for V_{X60} equations.

Symbols Used in Equations	
V_{X60}	volume of the unknown vessel at 60 °F
V_{S60}	volume of the standard vessel at 60 °F
$\rho_1, \rho_2, \ldots, \rho_N$	density of the water in the standard prover where ρ_1 is the density of the water for the first delivery, ρ_2 is the density of the water for the second delivery, and so on until all N deliveries are completed

$\Delta_1, \Delta_2,..., \Delta_N$	volume difference between water level and the reference mark on the standard where the subscripts 1, 2,..., N, represent each delivery as above. If the water level is below the reference line, Δ is negative. If the water level is above the reference line, Δ is positive. If the water level is at the reference line, Δ is zero NOTE: units must match volume units for the standard
$t_1, t_2, ..., t_N$	temperature of water for each delivery with the subscripts as above
α	coefficient of cubical expansion for the standard in units / °F
β	coefficient of cubical expansion for the prover in units / °F
t_x	temperature of the water in the filled unknown vessel in units °F
ρ_x	density of the water in the unknown vessel in g/cm^3

Note: Values for the density of water at the respective temperatures may be found in Table 9.8 (in NISTIR 6969). It may also be calculated using the formula given in GLP 10.

3.3 Pressure Corrections

Compute the pressure correction, P_{corr}, at each pressure that the prover was read, after any adjustments, by correcting for the compressibility of the water. The equation is:

$$P_{corr} = Scale\ reading\ @\ 100\ psig - scale\ reading\ @\ X\ psig$$
$$+ (water\ compressiblity\ factor)\left(\frac{100\ psig - X\ psig}{100}\right) \qquad \text{Eqn. 7}$$

where the water compressibility factor is 7.4 in^3 or 0.032 gal for a 100 gallon prover and 1.8 in^3 or 0.008 gal for a 25 gallon prover. The compressibility value is 0.00032 gal per gallon for 100 psig. The compressibility factor is given in both cubic inches and gallons so the proper unit can be selected depending upon the unit used for the scale readings.

Plot the pressure corrections. If the corrections versus the pressure are linear, make a straight line best fit of the data and interpolate to obtain the pressure corrections for any desired pressure. If the data is nonlinear, then perform a straight line interpolation between adjacent pressure readings to obtain pressure corrections at any desired intermediate pressures. Alternatively, a best fit curve can be drawn for the nonlinear data and the pressure corrections interpolated from the graph for intermediate pressures. See Appendix A for an example of the pressure corrections.

3.4 Prover Volume

LPG provers are generally adjusted to the nominal value using 100 psig as the reference pressure. At 100 psig, the prover reading will be lower by at least 0.032

gal/gal than it was at 0 psig due to the compressibility of water under pressure. Therefore, a prover of 100 gal should be set to read the V_{X60} value at 0 psig minus 7.4 in^3. (The water level on the prover's upper scale will be below the calculated V_{X60} reading when pressurized to 100 psig). The prover must not be set to read the V_{X60} value plus the compressibility factor at 0 psig because, in addition to compressibility of water, the prover expands as pressure is applied.

4   Measurement Assurance

4.1   Duplicate the process with a suitable check standard or have a suitable range of check standards for the laboratory. See SOP 17, SOP 20 and SOP 30. Plot the check standard volume and verify it is within established limits OR a *t*-test may be incorporated to check the observed value against an accepted value. The mean of the check standard observations is used to evaluate bias and drift over time. Check standard observations are used to calculate the standard deviation of the measurement process which contributes to the Type A uncertainty components.

4.2   If a standard deviation chart is used for measurement assurance, the standard deviation of each combination of Run 1 and Run 2 is calculated and the pooled (or average) standard deviation is used as the estimate of variability in the measurement process. Note: the pooled or average standard deviation over time will reflect varying conditions of test items that are submitted to the laboratory.

5   Assignment of Uncertainties

5.1   The limits of expanded uncertainty, U, include estimates of the standard uncertainty of the laboratory volumetric standards used, u_s, plus the uncertainty of measurement, u_m, at the 95 % level of confidence. See SOP 29 for the complete standard operating procedure for calculating the uncertainty.

5.2   The standard uncertainty for the standard, u_s, is obtained from the calibration report. The combined standard uncertainty, u_c, is used and not the expanded uncertainty, U, therefore the reported uncertainty for the standard will usually need to be divided by the coverage factor k.

Note: Fifteen is the maximum recommended number of deliveries from a laboratory standard to a prover under test to minimize calibration uncertainties to the levels identified previously.

5.3   Neck calibration uncertainty should be estimated based on the uncertainty of standards used, errors observed during calibration, and the repeatability of the neck calibration.

5.4 The standard deviation of the measurement process s_p, is taken from a control chart for an LPG check standard (not from charts for a refined fuel check standard) or standard deviation charts for LPG calibrations (See SOP 17, SOP 20 and SOP 30).

Other standard uncertainties usually included at this calibration level primarily include 1) uncertainties associated with the ability to read the meniscus, only part of which is included in the process variability due to parallax and visual capabilities, and 2) uncertainties associated with temperature corrections that include values for the cubical coefficient of expansion for the prover under test, the accuracy of temperature measurements, and factors associated with potential gradients in measuring the temperature in test measures or provers. Additional factors that might be included are: round robin data showing reproducibility, environmental variations over time, and bias or drift of the standard as noted in control charts.

5.5 To properly evaluate uncertainties and user requirements (tolerances), assessment of additional user uncertainties may be required by laboratory staff. Through proper use of documented laboratory and field procedures, additional uncertainty factors may be minimized to a level that does not contribute significantly to the previously described factors. Additional standard uncertainties in the calibration of field standards and their use in meter verification may include: how the prover level is established, how delivery and drain times are determined, the use of a proper "wet-down" prior to calibration or use, whether gravity drain is used during calibration or whether the volume of water is eliminated by pumping, the cleanliness of the prover and calibration medium, prover retention characteristics related to inside surface, contamination or corrosion, and total drain times, and possible air entrapment in the water. Systematic errors may be observed between laboratory calibration practices where a gravity drain is used and field use where the pumping system is used.

6 Report

6.1 Report results as described in SOP No. 1, Preparation of Calibration/Test Results, with the addition of the following:

For LPG provers, the total prover volume and uncertainty, reference temperature, material, coefficient of expansion (assumed or measured), any identifying markings, tolerances (if appropriate), laboratory temperature, water temperature, barometric pressure, relative humidity, out-of-tolerance conditions, and the total drain time from opening of the valve, including the 30 s drain after cessation of flow.

The report should also include a pressure correction table or chart, along with a note regarding possible differences in retention characteristics between water, the calibration medium, and LPG products.

Figure 1. LPG Prover Schematic

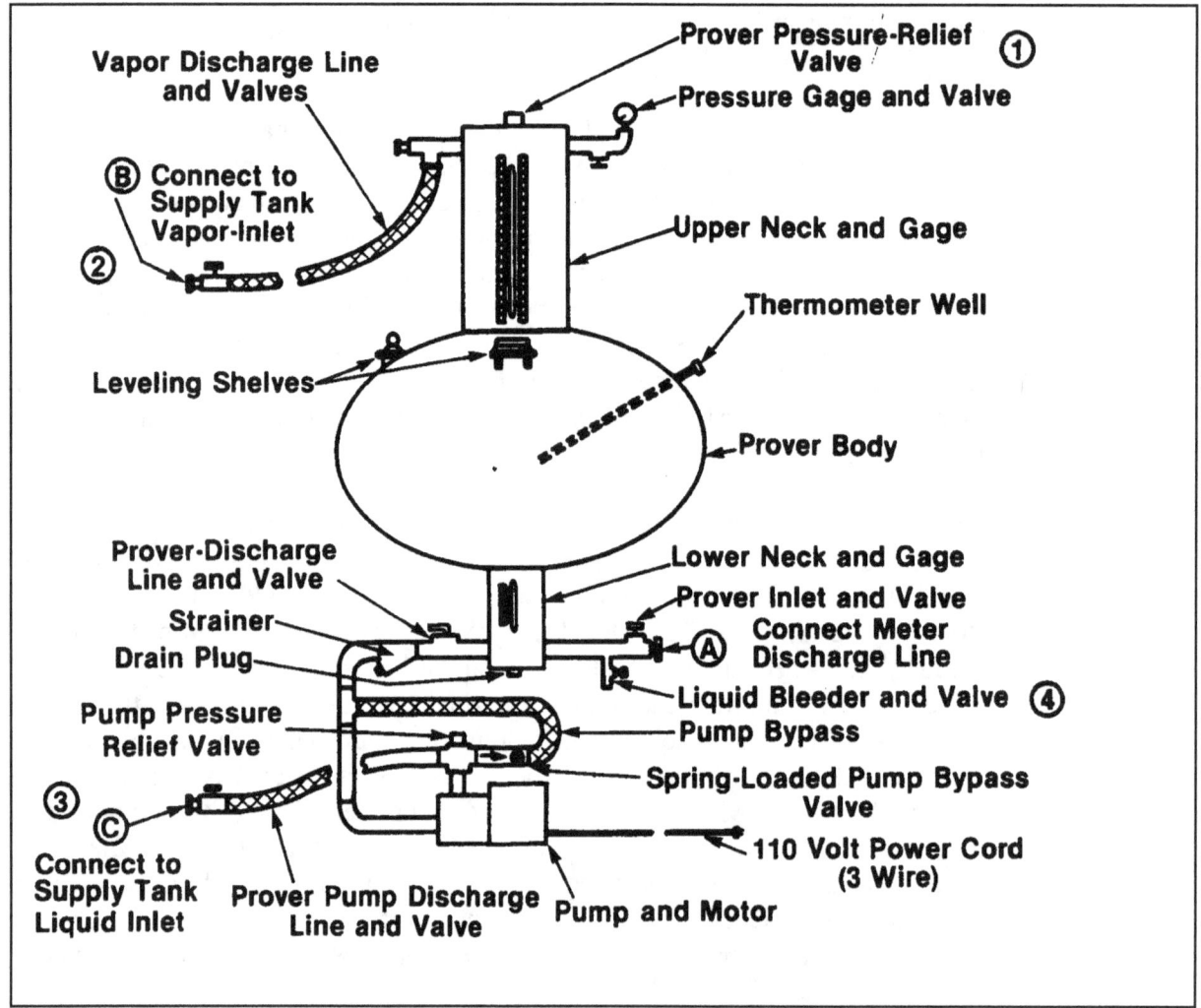

Appendix A

SOP No. 21
Recommended Standard Operations Procedure for
Calibration of LPG Provers

1   Compressibility of Water

The water compressibility factor is calculated based on an equation given in a paper by Frank E. Jones and Georgia L. Harris, "ITS-90 Density of Water Formulation for Volumetric Standards Calibration", as published in the Journal of Research of the National Institute of Standards and Technology, Vol. 97, No. 3, 1992.

$$\text{Compressibility factor} = \left(V_{NOM}\right)\left(231\right)\left(kT\right)\left(P\right)$$

where

V_{NOM}	Nominal volume of Prover in gal
kT	Thermal compressibility of water in $(atm)^{-1}$
P	Pressure in atm

And

$$kT = 50.83101x10^{-8} - 3.68293x10^{-9}t + 7.263725x10^{-11}t^2 - 6.597702x10^{-13}t^3 + 2.87767x10^{-15}t^4$$

where kT is in $(kPa)^{-1}$ and 1 atm = 101 325 Pa (exactly).

The thermal compressibility of water for the reference temperature 60 °F (15.56 °C) is $4.66264002 \times 10^{-9}$/kPa. LPG provers are calibrated at a nominal reference point at a specified temperature and pressure, typically 60 °F and 100 psig. A pressure of 100 psig correlates to 6.8 atmospheres. The following equation was used to calculate the compressibility factor:

Compressibility factor = $(V_{Nom})(231)(kT)(P)$

Compressibility factor = 100 gal (231 in³/gal)(4.72442x10⁻⁵/atm)(6.8 atm) = 7.4 in³ or 0.032 gal

1   Example of Pressure Corrections

The following example is printed from a spreadsheet program written by L. F. Eason, Metrologist (NC) to perform calculation for volume transfer. The program was modified by Georgia Harris (NIST) to provide calculations and corrections for LPG prover testing.

VOLUME TRANSFER CALIBRATION - METAL LPG PROVER

DATE: 04/09/91

COMPANY NAME:	Handbook 145 Example
ADDRESS:	12321 Some Street
CITY, STATE:	Anywhere, USA

NOMINAL VOLUME OF PROVER:	100 GALLONS
PROVER SERIAL NO:	100-234LP
MANUFACTURER:	Provers Unlimited
NUMBER OF DELIVERIES REQUIRED:	1 (0-15)

DATA FOR INDIVIDUAL STANDARD DELIVERIES

DROP NO	VOL (GAL)	METAL (MS/SS)	EXP COEF	TEMP DEG C	TEMP DEG F	WATER DENSITY	DELTA (CU IN)
1	100	stainless	0.00003	15.55	59.99	0.999013	0

TYPE OF METAL FOR THE UNKNOWN:	mild steel
COEFFICIENT OF EXPANSION:	0.0000186 /°F
WATER TEMPERATURE:	15.55 C
	59.99 F
WATER DENSITY:	0.999013 g/cm³
GAUGE AT 0 PSIG:	0.09 gal
GAUGE AT 50 PSIG:	0.02 gal
GAUGE AT 100 PSIG:	-0.05 gal
GAUGE AT 150 PSIG:	-0.10 gal
GAUGE AT 200 PSIG:	-0.16 gal
DELIVERY VOLUME AT 60 °F:	100.0000 V_{X60}, gallons
UNKNOWN PROVER'S ERROR AT 100 PSIG:	4.2 in³

ADJUSTMENT INSTRUCTIONS

To adjust prover to deliver exactly 100 gallons at 60 F, adjust scale to read -0.032 gallons or -7.4 in³, at 100 psig. Move to -0.032 gals, -7.4 in³, by 0.34 inches if 4" lower scale.

PROVER PRESSURE CORRECTIONS

	psig	P_{corr}	Prover Error	Prover Error	Volume of the Adjusted Prover (gal)
	0	-0.108	-0.108	-0.090	99.829
	50	-0.054	-0.054	-0.036	99.946
	100	0.000	0.000	0.018	100.000
	150	0.034	0.034	0.052	100.034
	200	0.078	0.078	0.096	100.078
			Adj.	Not Adj.	

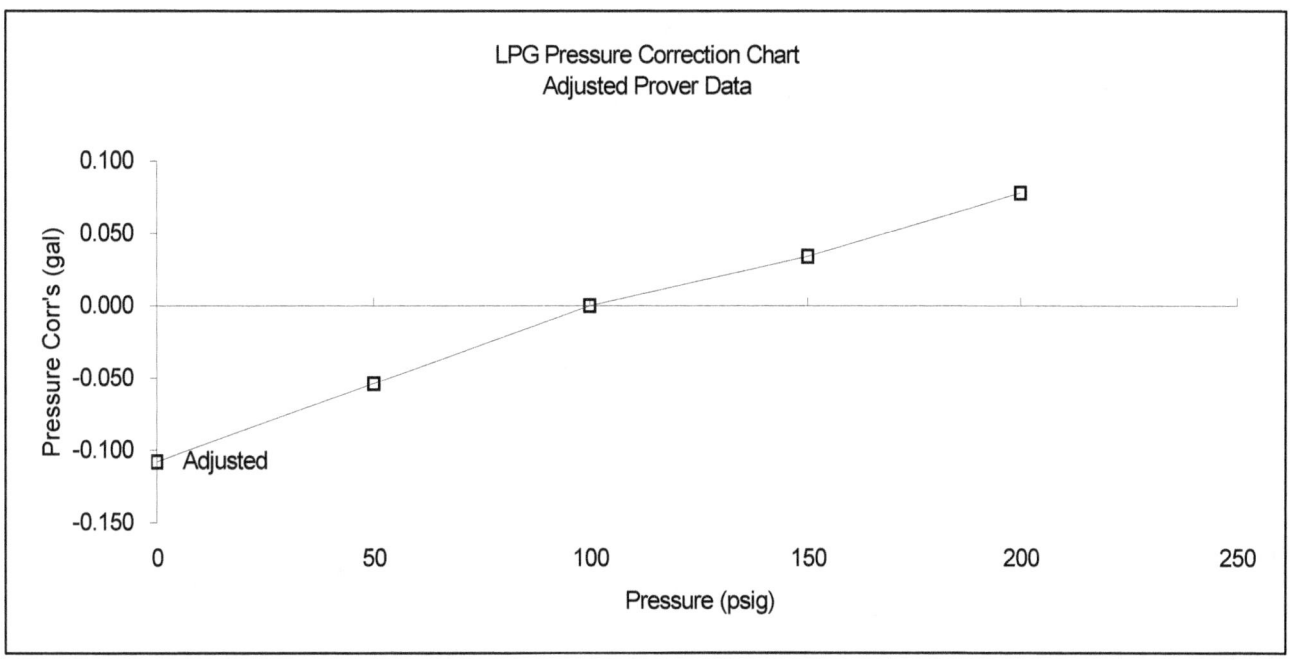

www.ingramcontent.com/pod-product-compliance
Lightning Source LLC
Chambersburg PA
CBHW081827170526
45167CB00007B/2743